Issues in Potable Reuse

The Viability of Augmenting Drinking Water Supplies With Reclaimed Water

Committee to Evaluate the Viability of Augmenting
Potable Water Supplies With Reclaimed Water

Water Science and Technology Board

Commission on Geosciences, Environment, and Resources

National Research Council

NATIONAL ACADEMY PRESS
Washington, D.C. 1998

National Academy Press • 2101 Constitution Avenue, NW • Washington, DC 20418

NOTICE: The project that is the subject of this report was approved by the Governing Board of the National Research Council, whose members are drawn from the councils of the National Academy of Sciences, the National Academy of Engineering, and the Institute of Medicine. The members of the committee responsible for the report were chosen for their special competencies and with regard for appropriate balance.

Support for this project was provided by the American Water Works Association Research Foundation, County Sanitation Districts of Los Angeles County, National Water Research Institute, Phoenix Water Services Department, San Diego County Water Authority, U.S. Bureau of Reclamation under grant nos. 1425-6-FG-81-07010 and 1425-6-FG-30-00740, U.S. Environmental Protection Agency under cooperative agreement no. CX 824340-01-0, and Water Environment Research Foundation.

Library of Congress Cataloging-in-Publication Data

Issues in potable reuse : the viability of augmenting drinking
water supplies with reclaimed water / Committee to Evaluate the
Viability of Augmenting Potable Water Supplies With Reclaimed
Water, Water Science and Technology Board, Commission on
Geosciences, Environment, and Resources, National Research Council.
     p. cm.
   Includes bibliographical references and index.
   ISBN 0-309-06416-3
   1. Water reuse. 2. Drinking water. I. National Research Council
(U.S.). Committee to Evaluate the Viability of Augmenting Potable
Water Supplies with Reclaimed Water.
   TD429 .I84 199   98-19686
   363.6'1—ddc21

Issues in Potable Reuse: The Viability of Augmenting Drinking Water Supplies With Reclaimed Water is available from the National Academy Press, 2101 Constitution Ave., N.W., Box 285, Washington, D.C. 20418 (1-800-624-6242; http://www.nap.edu).

Cover art by Y. David Chung. Chung is a graduate of the Corcoran School of Art in Washington, D.C. He has exhibited his work throughout the country, including at the Whitney Museum in New York, the Washington Project for the Arts in Washington, D.C., and the Williams College Museum of Art in Williamstown, Massachusetts.

COMMITTEE TO EVALUATE THE VIABILITY OF AUGMENTING POTABLE WATER SUPPLIES WITH RECLAIMED WATER

JAMES CROOK, *Chair*, Black and Veatch, Boston, Massachusetts *(from August 1996)*
RICHARD S. ENGELBRECHT, *Chair,* University of Illinois, Urbana-Champaign *(through August 1996)*
MARK M. BENJAMIN, University of Washington, Seattle
RICHARD J. BULL, Pacific Northwest National Laboratory, Richland, Washington
BRUCE A. FOWLER, University of Maryland, Baltimore
HERSCHEL E. GRIFFIN, San Diego State University, California
CHARLES N. HAAS, Drexel University, Philadelphia, Pennsylvania
CHRISTINE L. MOE, University of North Carolina, Chapel Hill
JOAN B. ROSE, University of South Florida, St. Petersburg
R. RHODES TRUSSELL, Montgomery Watson, Inc., Pasadena, California

Editor

DAVID A. DOBBS, Montpelier, Vermont

Staff

JACQUELINE A. MACDONALD, Study Director *(from July 1997)*
GARY D. KRAUSS, Study Director *(until July 1997)*
ELLEN A. DE GUZMAN, Senior Project Assistant

iii

WATER SCIENCE AND TECHNOLOGY BOARD

The National Academy of Sciences is a private, nonprofit, self-perpetuating society of distinguished scholars engaged in scientific and engineering research, dedicated to the furtherance of science and technology and to their use for the general welfare. Upon the authority of the charter granted to it by the Congress in 1863, the Academy has a mandate that requires it to advise the federal government on scientific and technical matters. Dr. Bruce M. Alberts is president of the National Academy of Sciences.

The National Academy of Engineering was established in 1964, under the charter of the National Academy of Sciences, as a parallel organization of outstanding engineers. It is autonomous in its administration and in the selection of its members, sharing with the National Academy of Sciences the responsibility for advising the federal government. The National Academy of Engineering also sponsors engineering programs aimed at meeting national needs, encourages education and research, and recognizes the superior achievements of engineers. Dr. William A. Wulf is president of the National Academy of Engineering.

The Institute of Medicine was established in 1970 by the National Academy of Sciences to secure the services of eminent members of appropriate professions in the examination of policy matters pertaining to the health of the public. The Institute acts under the responsibility given to the National Academy of Sciences by its congressional charter to be an adviser to the federal government and, upon its own initiative, to identify issues of medical care, research, and education. Dr. Kenneth I. Shine is president of the Institute of Medicine.

The National Research Council was organized by the National Academy of Sciences in 1916 to associate the broad community of science and technology with the Academy's purposes of furthering knowledge and advising the federal government. Functioning in accordance with general policies determined by the Academy, the Council has become the principal operating agency of both the National Academy of Sciences and the National Academy of Engineering in providing services to the government, the public, and the scientific and engineering communities. The Council is administered jointly by both Academies and the Institute of Medicine. Dr. Bruce M. Alberts and Dr. William A. Wulf are chairman and vice chairman, respectively, of the National Research Council.

Dedication

This book is dedicated to Dr. Richard Engelbrecht, who chaired the Committee to Evaluate the Viability of Augmenting Potable Water Supplies With Reclaimed Water from its inception until September 1, 1996, when he passed away.

Dr. Engelbrecht was a pioneer in environmental pollution control, particularly in issues related to water quality management and microbiological contamination of drinking water. A dedicated educator, he had been a professor of environmental engineering at the University of Illinois since 1954. He was also extremely active in the professional community, both nationally and internationally. Dr. Engelbrecht was elected to the National Academy of Engineering in 1976.

Dr. Engelbrecht found significant time between his teaching and research responsibilities and his professional involvements to volunteer his knowledge and expertise to further the work of the Water Science and Technology Board. A founding member of the board in 1982, Dr. Engelbrecht served on several of its study committees, including the Panel on Water Quality Criteria for Reuse, which in 1982 published the first independent review of potable reuse of reclaimed water. From 1988 until 1990, he chaired a committee that reviewed the U.S. Geological Survey's National Water Quality Assessment Program. He was a member of the Committee on Mexico City's Aquifer from 1992 until 1995 and the Panel on Sustainable Water and Sanitation Services for Megacities in the Developing World from 1995 until 1996.

Dr. Engelbrecht will long be remembered as one of the most influential figures in water quality and public health.

Preface

The National Research Council (NRC) first provided guidance on potable reuse of reclaimed wastewater in the report *Quality Criteria for Water Reuse,* developed in 1982 to provide input to an experimental program commissioned by Congress to study the wastewater-contaminated Potomac Estuary as a potential new water source for the District of Columbia. That report focused on the scientific questions concerning what quality criteria should be applied if a degraded water supply is to be used as a source of drinking water. At the time, only a few communities in the United States—notably Denver, Colorado; Los Angeles and Orange Counties, California; and Washington, D.C.—were considering water reuse to augment drinking water supplies.

The 1982 report produced a number of findings regarding the application of existing drinking regulations as applied to a reuse situation, the importance of treatment reliability and confirmatory data, and limitations concerning the identification, measurement, and long-term health risks from trace organic chemicals. The panel that wrote the report recommended that in the absence of an absolute, ideal water standard, the ability of a water reclamation facility to produce potable water should be judged—chemically, microbiologically, and toxicologically—in comparison with conventional drinking waters that are presumed to be safe. The panel suggested that all new projects develop an experimental pilot plant facility and recommended comprehensive toxicological testing to evaluate the potential health risks of unidentified trace organic chemicals. The report also "strongly endorse[d] the generally accepted concept that drinking water should be obtained from the best quality source available" and noted that "U.S. drinking water regulations were not established to judge the suitability of raw water supplies heavily contaminated with municipal and industrial wastewater." The report suggested that planners should consider "the much greater probability that adequately safe [reclaimed] water could be provided for short-term emergencies rather than for long-term use."

Since then, a number of factors have changed the way people think about drinking water and wastewater. The demand for water in some parts of the country is now so great that the best available sources have already been developed to their maximum extent, forcing municipalities to consider ways to creatively and inexpensively augment drinking water supplies. In addition, changes in water and wastewater treatment technologies have occurred since 1982, with advanced technologies such as advanced-oxidation (including ozone), membrane, and biofilm processes seeing increasing use in the United States and Europe. Finally, several public health concerns have surfaced over drinking water in general, and these concerns may affect the use of reclaimed water for drinking purposes. These include the control of *Cryptosporidium* and other potentially dangerous human pathogens, increased awareness of the potential dangers of disinfection by-products and other unidentified trace organic chemicals, and the problem of biological instability of wastewaters. When degraded water from whatever source is used, such issues complicate the challenge of meeting acceptable quality standards and ensuring that the water maintains its quality during distribution.

Since the 1982 NRC report, studies on the health implications of using reclaimed water for potable purposes have been completed at a number of projects in the United States. Many of the projects have used a comparative approach, testing both reclaimed and conventional drinking water sources. More advanced methods have been developed and tested for identifying organic chemicals and microbiological agents in reclaimed water, and some have involved whole-animal toxicological studies. A series of epidemiological studies was completed in Los Angeles County, where indirect potable reuse through ground water recharge has been practiced since 1962. None of the studies detected significant effects from chemical toxicants or infectious disease agents; all found the quality of highly treated reclaimed water as good as or better than the current drinking water sources for most or all measures of physical, chemical, and microbiological parameters.

However, limitations in methodology and testing have prevented many within the scientific and technical community from issuing absolute statements that planned potable reuse carries no adverse health-effect implications. Opponents of the use of reclaimed water for potable use point out that communities involved in the practice are subject only to existing drinking water standards that have been developed exclusively for natural sources of water. No national standards exist for the variety of contaminants (many of them poorly characterized) that may be present in potable water derived from treated municipal wastewater. Meanwhile, proponents have argued that the planned reuse situation is no different (and possibly safer because of tighter controls) than that faced by many communities using water supplies that receive significant upstream discharges of wastewater.

In conducting the review presented in this report, the Committee to Evaluate the Viability of Augmenting Potable Water Supplies With Reclaimed Water based its evaluation on published literature and the expertise of committee members and others consulted during this project. The committee used as its starting point the findings and recommendations of the 1982 NRC report. To gather further information, the committee hosted a two-day workshop in Irvine, California, with principal investigators and project managers of several of the potable reuse projects that have conducted analytical and health-effect studies.

I would personally like to thank my colleagues and fellow committee members for their cooperation, hard work, mutual respect, and enthusiasm. I would also like to thank the staff of the NRC's Water Science and Technology Board (WSTB), especially Gary Krauss and Jackie MacDonald, study directors, and Ellen de Guzman, senior project assistant. On behalf of the committee and WSTB, I would like to thank several people who provided important insight and contributed valuable information to the committee. These include Margie Nellor, Robert Baird, and Bill Yanko of the County Sanitation Districts of Los Angeles County; Michael Pereira of the Medical College of Ohio; Bill Lauer of the Denver Water Department; Bob Bastian of the Environmental Protection Agency; Joe Smith of the U.S. Bureau of Reclamation; Michael Wehner of Orange County Water District; and Adam Olivieri of the Western Consortium for Public Health. Finally, I would like to acknowledge the committee's deep appreciation to Richard Engelbrecht, the original chair of this committee, whose spirit and inspiration continued to help direct our efforts.

This report has been reviewed by individuals chosen for their diverse perspectives and technical expertise, in accordance with procedures approved by the NRC's Report Review Committee. The purpose of this independent review is to provide candid and critical comments that will assist the authors and the NRC in making the published report as sound as possible and to ensure that the report meets institutional standards for objectivity, evidence, and responsiveness to the study charge. The content of the review comments and draft manuscripts remain confidential to protect the integrity of the deliberative process. We wish to thank the following individuals for their participation in the review of this report:

Reviewers

Takashi Asano, University of California, Davis
Rebecca Calderon, Environmental Protection Agency
Russell Christman, University of North Carolina, Chapel Hill
Peter Gleick, Pacific Institute
Adel A. Mahmoud, Case Western Reserve University
Perry McCarty, Stanford University
Daniel Okun, University of North Carolina, Chapel Hill
Vernon Snoeyink, University of Illinois, Urbana-Champaign
Robert Spear, University of California, Berkeley

While the individuals listed above provided many constructive comments and suggestions, responsibility for the final content of this report rests solely with the authoring committee and the NRC.

JAMES CROOK, *Chair*
Committee to Evaluate the Viability of Augmenting
Potable Water Supplies With Reclaimed Water

Contents

Issues in Potable Reuse

Executive Summary

Growing populations and increasingly scarce new water sources have spurred a variety of water management measures over the last few decades, including the processing and reuse of water for many purposes. In a small but growing number of communities, these measures include the use of highly treated municipal wastewater to augment the raw water supply. This trend is motivated by need, but made possible by advances in treatment technology.

However, important questions remain regarding the levels of treatment, monitoring, and testing needed to ensure the safety of such "potable reuse." A 1982 National Research Council (NRC) report, *Water Quality Criteria for Reuse*, initially explored some of these questions. The significant advances, interest, and research in potable reuse since then, however, have spurred a reevaluation of these issues and this current report.

This study assesses the public health implications of using reclaimed water as a component of the potable water supply. It examines the different types of water reuse, discusses considerations for ensuring reliability and for evaluating the suitability of water sources augmented with treated wastewater, and seeks to identify future research needs regarding potable reuse safety testing and health effects

When considering potable reuse as an option for public water supplies, a critical distinction must be made between "direct" and "indirect" reuse. Direct potable reuse refers to the introduction of treated wastewater (after extensive processing beyond usual wastewater treatment) di-

1

rectly into a water distribution system without intervening storage. Direct use of reclaimed wastewater for human consumption, without the added protection provided by storage in the environment, is not currently a viable option for public water supplies. Instead, this report focuses on planned indirect potable reuse, which refers to the intentional augmentation of a community's raw water supply with treated municipal wastewater. The reclaimed water might be added to a water course, lake, water supply reservoir, or underground aquifer and then withdrawn downstream after mixing with the ambient water and undergoing modification by natural processes in the environment. The mix of reclaimed and ambient water is then subjected to conventional water treatment before entering the community's distribution system.

Planned indirect potable reuse cannot be considered in isolation from more general drinking water issues. Many communities currently use water sources of varying quality, including sources that receive significant upstream discharges of wastewater. In this sense, cities upstream of drinking water intakes are already providing water reclamation in their wastewater treatment facilities—for they treat the water, then release it into the raw water supply used by downstream communities. For example, more than two dozen major water utilities use water from rivers that receive wastewater discharges amounting to more than 50 percent of the stream flow during low flow conditions. Although most water systems using such raw water supplies meet current drinking water regulations, many of the concerns about planned, indirect potable reuse raised in this report also apply to these conventional water systems. The focus of this report, however, is planned indirect potable reuse of treated wastewater.

OVERALL CONCLUSIONS

The several indirect potable reuse projects currently operating in the United States generally produce reclaimed water that meets or exceeds the quality of the raw waters those systems would use otherwise, as measured by current standards. In some instances the reclaimed water meets or exceeds federal drinking water standards established by the Safe Drinking Water Act. Current potable reuse projects and studies have demonstrated the capability to produce reclaimed water of excellent measurable quality and to ensure system reliability. In communities using reclaimed water where analytical testing, toxicological testing, and epidemiological studies have been conducted, significant health risks have not been identified. This suggests that reclaimed water can likely be used safely to supplement raw water supplies that are subject to further treatment in a drinking water treatment plant. However, these projects raise some important questions: Can data from these projects safely be generalized to apply elsewhere? If not, what additional data are required? Do we know enough to establish criteria by which treated wastewater can be judged suitable for human consumption?

Our general conclusion is that planned, indirect potable reuse is a viable application of reclaimed water—but only when there is a careful, thorough, project-specific assessment that includes contaminant monitoring, health and safety testing, and system reliability evaluation. Potable reuse projects should include multiple, independent barriers that address a broad spectrum of microbiological and organic chemical contaminants. They should also conduct continuous toxicological monitoring if, as a result of the reclaimed water, the drinking water supply contains significant levels of organic contaminants of wastewater origin. Further, indirect potable reuse is an option of last resort. It should be adopted only if other measures—including other water sources, nonpotable reuse, and water conservation—have been evaluated and rejected as technically or economically infeasible.

It is important to recognize that although indirect potable reuse can be considered a viable option, many uncertainties are associated with assessing the potential health risks of drinking reclaimed water. These uncertainties are especially significant in toxicological and epidemiological studies. However, similar concerns also apply to the adequacy of these sciences for evaluating the safety of potable water from conventional sources, particularly the large number of sources already exposed to sewage contamination. These uncertainties are not an adequate reason for rejecting indirect potable reuse because the best available current information suggests that the risks from indirect potable reuse projects are comparable to or less than the risks associated with many conventional supplies.

That said, however, the intentional reuse of treated wastewater raises issues that must be addressed to ensure protection of public health. Drinking water standards cover only a limited number of contaminants. They are intended for water obtained from conventional, relatively uncontaminated sources of fresh water, not for reclaimed water, and therefore cannot be relied on as the sole standard of safety. The requirements for indirect potable reuse systems thus should exceed the requirements that apply to conventional drinking water treatment facilities.

The major recommendation of this report is that water agencies considering potable reuse fully evaluate the potential public health impacts from the microbial pathogens and chemical contaminants found or likely to be found in treated wastewater through special microbiological, chemical, toxicological, and epidemiological studies, monitoring programs, risk assessments, and system reliability assess-

ments. This report provides guidelines and suggestions regarding how such evaluations should be carried out. Thorough evaluation of the risks of a proposed potable reuse project, in addition to full consideration of other options for potable water supply augmentation, is essential for a sound decision about whether the project is viable.

CHEMICAL CONTAMINANTS IN REUSE SYSTEMS

Municipal wastewater contains chemical contaminants of three sorts: (1) inorganic chemicals and natural organic matter that are naturally present in the potable water supply; (2) chemicals created by industrial, commercial, and other human activities in the wastewater service area; and (3) chemicals that are added or generated during water and wastewater treatment and distribution processes. Any project designed to reclaim and reuse such water to augment drinking supplies must adequately account for all three categories of contaminants.

The organic chemicals in a wastewater present one of the most difficult challenges a public health engineer or scientist faces in considering potable reuse. The challenge arises from the large number of compounds that may be present, the inability to analyze for all of them, and the lack of toxicity information for many of the compounds. Efforts to account for the total mass of organic carbon in water are further frustrated by the fact that the bulk of this material is aquatic humus, which varies slightly in structure and composition from one molecule to the next and cannot be identified like conventional organic compounds. These challenges are not unique to potable reuse systems. In fact, the most protected water supplies are those for which the smallest fraction of the organic material can be identified. For potable reuse systems, however, anthropogenic organic compounds pose the greatest concern and should be the major focus of monitoring and control efforts.

The following recommendations suggest several important guidelines to account for chemical contaminants of potential concern in potable reuse systems:

• **The research community should study in more detail the organic chemical composition of wastewater and how it is affected by treatment, dilution, soil interaction, and injection into aquifers.** The composition of wastewater and the fate of the organic compounds it contains need to be better understood to increase the certainty that health risks of reclaimed water have been adequately controlled through treatment and storage in the environment.

• **Proposed potable reuse projects should include documentation of all major chemical inputs from household, industrial, and agricul-**

tural sources. Reuse project managers can estimate household chemical inputs on a per capita basis from general information about domestic wastewaters. Project managers should undertake a major effort to quantify the inputs of industrial chemicals, paying special attention to chemicals of greatest health concern.

• **Stringent industrial pretreatment and pollutant source control programs should be used to reduce the risk from the wide variety of synthetic organic chemicals (SOCs) that may be present in municipal wastewater and consequently in reclaimed water.** Guidelines for developing lists of wastewater-derived SOCs that should be controlled through industrial pretreatment and source control should be prepared by the Environmental Protection Agency (EPA) and modified for local use. Potable reuse operations should include a program to monitor for these chemicals in the treated effluent, tracking those that occasionally occur at measurable concentration with greater frequency than those that do not.

• **Although Safe Drinking Water Act regulations cannot alone ensure the safety of drinking water produced from treated wastewater, potable reuse projects must nonetheless bring contaminant concentrations within those regulations' guidelines.** Potable reuse projects can manage regulated contaminants by a combination of secondary or tertiary wastewater treatment processes, dilution or removal in the receiving water, and removal in the drinking water treatment plant.

• **The risks posed by unidentifiable or unknown contaminants in reuse systems should be managed by a combination of reducing concentrations of general contaminant classes, such as total organic carbon, and conducting toxicological studies of the water.** Reducing the level of organic matter to the lowest practical concentration will reduce but not necessarily eliminate the need for toxicological studies and monitoring. The nature of the organic carbon in the water will influence what the appropriate total organic carbon limit should be. This judgment should be made by local regulators, integrating all the available information concerning a specific project.

• **The research community should determine whether chlorination of wastewater creates harmful levels of unique disinfection by-products that might pose concerns in potable reuse systems.** Whether reclaimed water forms significantly different by-products than natural waters upon disinfection is not yet clear.

• **Finally, every reuse project should have a rigorous and regularly updated monitoring system to ensure the safety of the product water.** This program should be updated periodically as inputs to the system change or as its results reveal areas of weakness. Pretreatment requirements, wastewater treatment processes, and/or monitoring requirements may need to be modified to protect public health from exposure to specific chemical contaminants.

MICROBIAL CONTAMINANTS IN REUSE SYSTEMS

Microbial contaminants in reclaimed water include bacteria, viruses, and protozoan parasites. Even though classic waterborne bacterial diseases such as dysentery, typhoid, and cholera have dramatically decreased in the United States, *Campylobacter*, nontyphoid *Salmonella*, and pathogenic *Escherichia coli* still cause a significant number of illnesses, and new emerging diseases pose potentially significant health risks.

Historically, coliforms, which serve as an adequate treatment indicator or "marker" for many bacterial pathogens of concern, have been used as general indicators of the levels of microbial contamination in drinking water. Today, however, most outbreaks of waterborne disease in the United States are caused by protozoan and viral pathogens in waters that meet coliform standards. Yet few drinking water systems, either conventional ones or those involving potable reuse, monitor for the full range of such pathogens, and little information exists regarding the efficacy of water and wastewater treatment processes in removing them. In addition, wastewater may contain a number of newly recognized or "emerging" waterborne enteric pathogens or potential pathogens.

To ensure the safety of drinking water, planners, regulators, and operators of potable reuse systems should take steps to further reduce the various existing and potential health risks posed by these microbial contaminants:

- **Potable reuse systems should continue to employ strong chemical disinfection processes to inactivate microbial contaminants even if they also use physical treatment systems to remove these contaminants.** Some new membrane water filtration systems can almost completely remove microbial pathogens of all kinds, but experience with them is not yet adequate to depend on them alone for protection against the serious risks posed by these pathogens. Therefore, strong chemical disinfectants, such as ozone or free chlorine, should also be used, even in systems that include membrane filters.
- **Managers of current and future potable reuse facilities should assess and report the effectiveness of their treatment processes in removing microbial pathogens so that industry and regulators can develop operational guidelines and standards.** Reuse project managers should provide data on number of barriers, microbial reduction performance, and reliability or variation.

• **The potable reuse industry and the research community should establish the performance and reliability of individual barriers to microorganisms within treatment trains and should develop performance goals appropriate for planned potable reuse.** Existing microbial standards for drinking water systems assume that the water source is natural surface or ground water. Treatment standards and goals more appropriate for potable reuse projects need to be developed.

METHODS FOR ASSESSING HEALTH RISKS OF RECLAIMED WATER

Any effort to augment potable water supplies with reclaimed water must include an evaluation of the potential health risks. Such assessment is complicated by several factors, including uncertainties about the potential contaminants and contaminant combinations that may be found in reclaimed water and about the human health effects those contaminants may cause. Any such effort must evaluate health risks from both microbial and chemical contaminants.

Microbiological Methods and Risk Assessment

The lack of information nationwide on the levels of viral and protozoan pathogens in all waters and the efficacy of both conventional water treatment and wastewater treatment for water reclamation in removing those pathogens poses challenges to estimating risks from microbial contamination in potable reuse systems. The Information Collection Rule promulgated in 1996 by the EPA should help provide the exposure data needed for more effective risk assessments, but additional steps are needed to improve methods for assessing risks posed by microbial pathogens in water reuse projects:

• **Potable reuse project managers should consider using some of the newer analytical methods, such as biomolecular methods, as well as additional indicator microorganisms, such as *Clostridium perfringens* and the F-specific coliphage virus, to screen drinking water sources derived from treated wastewaters.** These screening methods should cover some of the gaps in analysis left by the bacterial and cell culture methods currently used for detecting pathogens in water. Additional research should be sponsored to improve methods for detecting emerging pathogens in environmental samples.
• **The EPA should include data on the concentrations of waterborne pathogens in source water in the new Drinking Water National Contaminant Occurrence Data Base and should develop better data on**

the reductions of pathogens accomplished by various levels of treatment. The lack of monitoring data for evaluating exposure remains the greatest single barrier to adequate risk assessment for microbial pathogens.

• The research community should conduct more research to document removal rates of relevant protozoa by natural environmental processes. Indirect potable reuse projects may rely on dilution in the environment, die-off in ambient waters, and removal by soil infiltration to reduce concentrations of microbial pathogens. While reductions of bacteria and virus concentrations in natural environments have been well documented, information on protozoa survival in ambient waters remains inadequate.

• Risk estimates should consider the effects pathogens may have on sensitive populations and the potential for secondary spread of infectious disease within a community. This precaution is necessary to prevent pathogens from infecting sensitive populations (the elderly or very young, or those with suppressed immune systems) in whom mortality may be high and from whom diseases might spread to others.

• The research community should carry out further studies to document the removal of pathogens of all types by membrane filtration systems. Certain membrane filtration systems show the potential for nearly complete removal of pathogens. More research is needed to demonstrate the suitability of these systems for potable reuse applications and to develop monitoring methods capable of continuously assessing system performance.

Chemical Risk Assessment

Because of the uncertainty of the organic chemical composition of reclaimed water, toxicological testing should be the primary component of chemical risk assessments of potable reuse systems. However, recent experience and research have shown that the conventional toxicological testing strategies developed in the food and drug industries, as well as the similar testing protocols recommended by the NRC in its 1982 report, are not adequate for evaluating risks from the complex chemical mixtures found in reclaimed wastewater. These testing protocols, which stress the use of concentrates of representative organic chemicals in both *in vitro* (cell culture) and *in vivo* (whole-animal) tests, have several critical limitations. These limitations include uncertainty as to whether the concentrates used for testing are truly representative of those in the wastewater; higher than expected occurrences of false negative results; long lag times between sample collection and the availability of results; difficulty in tracing results to particular constituents; and lack of suitability for continuous monitoring. In addition, a truly thorough application of the NRC protocol, which would involve extensive testing of concentrates on live animals, is both expensive and time-consuming.

Given these complications, in waters where toxicological testing appears to be important for determining health risks, emphasis should be placed on live animal test systems that are capable of expressing a wide variety of toxicological effects. Chapter 5 presents a suggested approach using fish populations in unconcentrated treated wastewater.

Further, toxicological testing standards for reclaimed water should be supplemented by strict regulation of the processes for "manufacturing" the water. Regulators should review the processes for manufacturing the reclaimed water (that is, the treatment systems and environmental storage employed) on a plant-by-plant basis.

HEALTH-EFFECT STUDIES OF REUSE SYSTEMS

The few studies that have examined the health effects of drinking reclaimed water suggest that the current approaches to safety testing of reclaimed water, derived mainly from consumer product testing protocols, are inadequate for evaluating reclaimed water and should be replaced by a more appropriate method. Even a brief look at these studies makes clear the need for a new approach.

Toxicology Studies

This report includes a review of six planned potable reuse projects that tried to analyze and compare the toxicological properties of reclaimed water to those of the communities' current drinking water supplies. In most of the six studies, testing was limited to assessing whether the water caused genetic mutations in bacterial systems. Some studies also used *in vitro* systems derived from mammalian cells, and two projects also used chronic studies in live mammal systems. Only two studies, carried out in Denver and Tampa, addressed a broad range of toxicological concerns. Those studies suggested that no adverse health effects should be anticipated from the use of Denver's or Tampa's reclaimed water as a source of potable drinking water. However, these studies, drawn from two discrete points in time and conducted only at a pilot plant level of effort, provide a very limited database from which to extrapolate to other locations and times.

Overall, the intent of toxicological testing can be grouped into (1) chemical screening and identification studies; (2) surveys to determine genetic mutation potentials; and (3) integrated toxicological testing. In theory, all three stages will be applied when needed. In practice, the application of the third and crucial stage, that of integrated testing, has been both uneven and impractically expensive.

Screening studies merely identify chemicals that may be causing mutations; the mutagenic activity may or may not cause health problems. Gauging the actual health risk such chemicals pose requires test systems that can more directly measure a complete range of health hazards and can define dose-response relationships that allow an estimate of the risks associated with various levels of exposure. The most common method to accomplish this goal in conventional product safety testing is to test the chemicals or contaminants in question at doses approaching the maximally tolerated dose so as to establish the margin between environmental levels and those that produce adverse effects. For reclaimed water, however, the high cost and methodological problems inherent in this approach make it both unreliable and inefficient.

Accordingly, a new, alternative testing approach, such as one using fish in source water, should be developed to allow continuous toxicological testing of reclaimed water at reasonable cost. The system, an example of which is discussed in detail in Chapter 5, should employ a baseline screening test using a whole-animal rather than an *in vitro* approach and should be modified as results and research suggest improvements. The tests should use water samples at ambient concentrations in order to reduce the uncertainty and high costs of using concentrates. Any losses in sensitivity from not using high doses should be offset by the increased statistical power brought by using larger numbers of whole-animal test subjects, such as fish.

Research efforts should investigate the qualitative and quantitative relationships among responses in whole-animal test species, such as fish, and adverse health effects in humans. *In vitro* short-term testing using concentrations of chemicals should be confined to qualitative evaluations of particular toxicological effects found in the product water in order to identify potential sources of contaminants and to quickly guide remedial actions.

For any toxicological test used for reclaimed water, a clear decision path should be followed. Testing should be conducted on live animals for a significant period of their life span. If an effect is observed, risk should be estimated using state-of-the-art knowledge about the relative sensitivity of the animal and human systems, and, if warranted, further defined by more research. This decision path is quite workable if the underlying basis of the biological response in question is understood (for example, endocrine disruption). For some health outcomes, such as carcinogenesis, the mechanism is less well understood, and an observed effect may have to be accepted as implying an impact on human health.

The need for toxicological testing of water is inversely related to

how well the water's chemical composition has been characterized. If a water contains very few or very low concentrations of chemicals or chemical groups of concern, the need for toxicological characterization of the water may be substantially reduced. Conversely, if a large fraction or high concentrations of potentially hazardous and toxicologically uncharacterized organic chemicals are present, toxicological testing will provide an additional assurance of safety.

Epidemiologic Studies

Numerous epidemiologic studies have examined the relationship between contaminants in drinking water and health problems. However, only three such studies apply to potable reuse of reclaimed water, and only one set of epidemiological studies (Los Angeles County) has been conducted in a setting that can be generalized to apply to other communities. These studies have used an ecologic approach, which is appropriate as an initial step when health risks are unknown or poorly documented, but negative results from such studies do not prove the safety of the water in question. These studies can only be considered as preliminary examinations of the risks of exposure to reclaimed water. Epidemiological data that can be confidently applied to the potable use of reclaimed water are lacking. Filling that knowledge gap would aid planning and help ensure the safety of such projects.

The 1982 NRC report on potable reuse concluded that "unless epidemiological methodology is improved, it is doubtful whether it can be used to evaluate the potential carcinogenic risk of drinking reused water" but recommended monitoring for acute waterborne diseases. Since that report, at least 17 large epidemiological studies (using several designs) have examined the association between chlorinated surface water and cancer, and two large cohort studies have examined the risk of endemic waterborne disease due to infectious agents. These studies have greatly increased our experience with exposure assessment, and outcome measurements in this area could be used to help design future epidemiologic studies of reclaimed water.

Therefore, epidemiologic studies should be conducted at the national level using alternative study designs and more sophisticated methods of exposure assessment and outcome measurement to evaluate the potential health risks associated with reclaimed water. Ecologic studies should be conducted for water reuse systems using ground water and surface water in areas with low population mobility. Case-control studies or retrospective cohort studies should be undertaken to provide information on health outcomes and exposure on an individual level while controlling for other important risk factors.

RELIABILITY AND QUALITY ASSURANCE ISSUES FOR REUSE SYSTEMS

To ensure that any temporary weaknesses in the treatment process or water quality are promptly detected and corrected, potable reuse systems should provide multiple barriers to contamination and should monitor both water quality and potential health effects of substandard water, according to the following guidelines:

• **Potable water reuse systems should employ multiple, independent barriers to contaminants, and the barriers should be evaluated both individually and together for their effectiveness in removing each contaminant of concern.** Further, the cumulative capability of all barriers to accomplish removal should be evaluated, and this evaluation should consider the levels of the contaminant in the source water.

• **Barriers for microbiological contaminants should be more robust than those for forms of contamination posing less acute dangers.** The number of barriers must be sufficient to protect the public from exposure to microbial pathogens in case one of the barriers fails.

• **Because performance of wastewater treatment processes may vary, such systems should employ quantitative reliability assessments to gauge the probability of contaminant breakthrough among individual unit processes.** "Sentinel parameters," indicators of treatment process malfunctions that are readily measurable on a rapid (even instantaneous) basis and that correlate well with high contaminant breakthrough, should be used to monitor critical processes that must be kept under tight control.

• **Utilities using surface waters or aquifers as environmental buffers should take care to prevent "short-circuiting," a process by which treated wastewater influent either fails to fully mix with the ambient water or moves through the system to the drinking water intake faster than expected.** In addition, the buffer's expected retention time should be long enough—probably 6 to 12 months, as outlined in recently proposed California regulations—to give the buffer time to provide additional contaminant removal. Such a lag time also allows public health authorities to take action in the event that unanticipated problems arise in the water reclamation system.

• **Potable reuse operations should have alternative means for disposing of the reclaimed water in the event that it does not meet required standards.** Such alternative disposal routes protect the environmental buffer from contamination.

• **Every community using reclaimed waters as drinking water should implement well-coordinated public health surveillance systems** to document and possibly provide early warning of any adverse health events associated with the ingestion of reclaimed water. Any such surveillance system must be jointly planned and operated by the health, water, and wastewater departments and should identify key individuals in each agency to coordinate planning and rehearse emergency procedures. Further, appropriate interested consumer groups should be involved with and informed about the public health surveillance plan and its purpose.

- Finally, operators of water reclamation facilities should receive training regarding the principles of operation of advanced treatment processes, the pathogenic organisms likely to be found in wastewaters, and the relative effectiveness of the various treatment processes in reducing contaminant concentrations. Operators of such facilities need training beyond that typically provided to operators of conventional water and wastewater treatment systems.

1

Reclaiming Wastewater: An Overview

Growing urbanized populations and increasing constraints on the development of new water sources have spurred a variety of measures to conserve and reuse water over the last two or three decades. As part of this trend, some municipalities have begun to reuse municipal wastewater for nonpotable water needs, such as irrigation of parks and golf courses. And a small but increasing number of municipalities are augmenting or considering augmenting the general water supply (potable and nonpotable) with highly treated municipal wastewater.

These "potable reuse" projects are made possible by improved treatment technology that can turn municipal wastewater into reclaimed water that meets standards established by the Safe Water Drinking Act. However, questions remain regarding how much treatment and how much testing are necessary to protect human health when reclaimed water is used for potable purposes. Some public health and engineering professionals object in principle to the reuse of wastewater for potable purposes, because standard public health philosophy and engineering practice call for using the purest source possible for drinking water. Others worry that current techniques might not detect all the microbial and chemical contaminants that may be present in reclaimed water. Several states have issued regulations pertaining to potable reuse of municipal wastewater, but these regulations offer conflicting guidance on whether potable reuse is acceptable and, when it is acceptable, what safeguards should be in place.

This report assesses the health effects and safety of using reclaimed water as a sole source or as a component of the potable water supply. The report was prepared by the Committee to Evaluate the Viability of Augmenting Potable Water Supplies With Reclaimed Water, which was appointed by the National Research Council (NRC) to evaluate issues associated with potable reuse of municipal wastewater. The committee members were appointed based on their widely recognized expertise in municipal water supply, wastewater reclamation and reuse, and public health. In its evaluation, the committee considered the following questions:

- What are the appropriate definitions of water reuse? What distinguishes indirect from direct reuse?
- What are the considerations for ensuring reliability and for evaluating the suitability of a water source augmented with treated wastewater?
- Given the recent health-effect studies that have been conducted, what further research is required?

The committee based its evaluation on published literature and the expertise of committee members and others consulted during this project. The committee used as its starting point the findings and recommendations of a 1982 NRC committee that examined quality criteria that should be applied when a degraded water supply is used as a drinking water source (see Box 1-1). As part of its information gathering effort, the committee hosted a two-day workshop in Irvine, California, featuring principal investigators and project managers of several of the potable reuse projects that have conducted analytical and health-effect studies.

The committee views the planned use of reclaimed water to augment potable water supplies as a solution of last resort, to be adopted only when all other alternatives for nonpotable reuse, conservation, and demand management have been evaluated and rejected as technically or economically infeasible. This report should help communities considering potable reuse make decisions that will protect the populations they serve. Some of the issues relate to similar concerns for drinking water sources that receive incidental or unplanned upstream wastewater discharges.

This chapter describes the history of potable reuse of municipal wastewater, defines the different types of potable reuse, provides an overview of wastewater treatment technologies applicable to potable reuse projects, and describes existing federal guidelines and state regulations covering potable reuse. Chapter 2 describes the chemical contaminants found in wastewater, treatments aimed at reducing them, and issues re-

BOX 1-1
Results of the 1982 NRC Study of Quality Criteria for Water Reuse

In 1982, the National Research Council issued a report titled *Quality Criteria for Water Reuse* (National Research Council, 1982). The report was developed to provide input to an experimental program commissioned by Congress to study the wastewater-contaminated Potomac Estuary as a potential new water source for the District of Columbia (National Research Council, 1982). The focus of the 1982 report was on the scientific questions concerning the quality criteria that should be applied if a degraded water supply is used as a source of drinking water. At the time, very few communities in the United States, aside from Denver, Los Angeles, Washington, D.C., and Orange County, California, were considering water reuse to augment drinking water supplies.

The report concluded that the most practical way to judge the potential health hazards of reclaimed water is to compare it with conventional supplies, which have risks, if any, that are presumed to be acceptable. Initially, conventional water supplies and reclaimed water should be compared on the basis of identifiable individual compounds and microbiological organisms. The results of these tests would influence the need to proceed with additional testing, because reclaimed water that failed such a comparison would be rejected as not being as suitable as a conventional supply. Because of the practical impossibility of identifying and testing all of the individual compounds present in reclaimed water, the report recommended testing of mixtures of chemicals. It also recommended that the mixtures be concentrated to increase the sensitivity of the tests.

The report recommended that toxicological comparisons between reclaimed and conventional water be based on the outcomes of a series of tiered tests designed to provide information on the relative toxicities of the concentrates from the two water supplies. Phase 1 tests would include *in vitro* assessments of mutagenic and carcinogenic potential by means of microbial and mammalian cell mutation and *in vivo* evaluations of acute and short-term subchronic toxicity, teratogenicity (birth defects), and clastogenicity (the production of chromosomal abnormalities). Phase 2 tests would include a longer term (90-day) subchronic study and a test for reproductive toxicity. Phase 3 would consist of a chronic lifetime feeding study.

The report concluded that, depending on the results of the various comparative test phases, a judgment could be reached that reclaimed water is as safe as, more safe than, or less safe than a conventional water supply. The final decision to use treated wastewater for potable purposes or for food processing would only be made after a careful evaluation of potential health effects, treatment reliability, cost, necessity, and public acceptance. Still, the report "strongly endorse[d] the generally accepted concept that drinking water should be obtained from the best quality source available" and noted that "U.S. drinking water regulations were not established to judge the suitability of raw water supplies heavily contaminated with municipal and industrial wastewater." The report suggested that planners should consider "the much greater probability that adequately safe [reclaimed] water could be provided for short-term emergencies rather than for long-term use."

lated to analytical methods for measuring water quality. Chapter 3 examines similar concerns related to microbial contaminants. Chapter 4 discusses methodological issues for conducting microbiological analysis, risk analysis, toxicological safety testing, and epidemiological studies. Chapter 5 reviews the health-related studies conducted by selected potable reuse projects. And Chapter 6 evaluates reliability and quality assurance issues for potable reuse projects.

SELECTION OF DRINKING WATER SOURCES

Some public health authorities have been reluctant to allow or support the planned augmentation of water supplies with reclaimed municipal wastewater under any circumstances, subscribing to the maxim that only natural water derived from the most protected source should be used as a raw drinking water supply. This maxim has guided the selection of potable water supplies for more than 150 years. It was affirmed in the 1974 draft of the National Interim Primary Drinking Water Regulations, which states, "Production of water that poses no threat to the consumer's health depends on continuous protection. Because of human frailties associated with protection, priority should be given to selection of the purest source. Polluted sources should not be used unless other sources are economically unavailable" (U.S. EPA, 1975).

This principle was derived from earlier public health practices developed when understanding of drinking water contaminants was limited and when natural processes (such as dilution in rivers and natural filtration by soils), rather than technology, were relied upon to produce suitable drinking water. It is also derived from a time when the U.S. population was smaller, and our concern about protecting the environment from the impact of human-made impoundments less formalized, and when pristine water supplies were more available than they are today.

While a pristine drinking water source is still the ideal sought by most municipalities, the U.S. population has expanded, so that many large cities take water from sources that are exposed to sewage contamination. When these supplies were originally developed, the only health threats perceived were attributable to microbiological vectors of infectious disease. These vectors would be attenuated during flow in rivers and then easily eliminated with conventional water treatment processes such as coagulation, filtration, and disinfection. Such water supplies were generally less costly and more easily developed than higher quality upland supplies or underground sources. Today, however, most of these supply waters receive treated wastewaters from other communities upstream. Thus, cities such as Philadelphia, Cincinnati, and New Orleans, which draw water from the Delaware, Ohio, and Mississippi rivers, respectively, are already practicing unplanned indirect potable reuse of municipal wastewater. In fact, more than two dozen major water utilities, serving populations from 25,000 to 2 million people, draw from rivers in which the total wastewater discharge accounts for more than 50 percent of stream flow during low flow conditions (Swayne et al., 1980).

Much of the impetus for water reuse comes from municipal utilities in the arid western United States. Many communities there already use a variety of measures to offset the rising costs of importing water long distances. Moving water entails satisfying a large number of environmental and health laws and permits, as well as the corresponding interests of competing users and local, state, tribal, and federal jurisdictions.

As high-quality water sources become scarcer and populations in arid regions grow, the phrase "economically unavailable" has taken on new significance. Communities looking for new water sources must examine a number of options, including water conservation, nonpotable reuse, and investing more money in the treatment of water supplies that are of poorer quality but more readily available. Most communities will readily pay a premium to obtain a pristine supply for their drinking water. But the premiums required get bigger each year, particularly in areas where water is already scarce.

POTABLE REUSE AND CURRENT DRINKING WATER STANDARDS

Much of the objection to planned potable reuse of wastewater arises from a discussion of whether drinking water standards are adequate to ensure the safety of all waters regardless of source. Some argue that drinking water standards apply only to—and were designed only for—waters derived from relatively pristine sources. Although this argument has a long-standing basis in normal sanitation practice, it is becoming more difficult to determine what is the best available water source. Water sources in the United States vary from protected, pristine watersheds to waters that have received numerous discharges of various wastes, as illustrated in Figure 1-1. Highly treated wastewater does not differ substantially from some sources already being used as water supplies.

Because of the continuing degradation of raw water supplies in the United States and increased public concern about water quality, federal drinking water regulations, which in 1925 addressed only a handful of contaminants and applied only to municipalities that provided water to interstate carriers (such as buses, trains, and ships), now address nearly 100 contaminants and apply to all community water systems serving 25 people or more. The role of these drinking water standards should be evaluated against the continuum of available source waters.

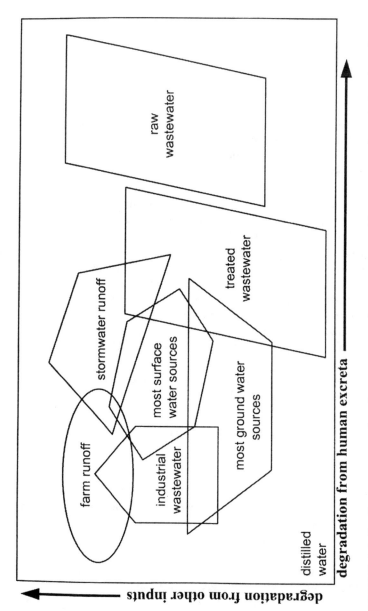

FIGURE 1-1 Quality spectrum of various waters and wastewaters with respect to degradation from human excreta and other materials.

Drinking water standards' main function is to provide a benchmark for unacceptable risk from selected contaminants for which adequate health information exists. Up to a point, increasing the number of standards increases the confidence that a particular water supply is not contaminated by harmful chemicals or pathogens. However, the standards cannot guarantee that the water poses no health hazard. Modern analytical methods detect fewer than 10 percent of the organic chemicals typically present in a water (Ding et al., 1996). Further, drinking water standards exist only for a relatively small percentage of the possible chemical contaminants. In addition, these standards do not currently require monitoring for specific microbiological contaminants, but only for coliform bacteria, which merely indicate the possible presence of microbial pathogens—and only a fraction of microbial pathogens at that. Creating more standards, therefore, does not ensure the safety of drinking water, because as more chemical contaminants and pathogenic organisms are discovered the possibilities become almost infinite in scope.

In summary, caution is required when evaluating whether compliance with drinking water standards—or proposed or hypothetical additional standards—will ensure a water source is safe. As a water source comes to include (intentionally or not) increasing amounts of wastewater, a drinking water utility must become increasingly knowledgeable about contaminant inputs into the wastewater. The utility might identify potential contaminants of concern by surveying the industrial inputs into the wastewater, examining the wastewater for chemical constituents broader than those represented by drinking water standards, and/or using toxicological testing methods to ensure that the product water does not contain substantial concentrations of chemicals whose toxicological properties have not been established.

TYPES OF WATER REUSE

When discussing the reuse of treated municipal wastewater for potable purposes, it is useful to distinguish between "indirect" and "direct" potable reuse and between "unplanned" and "planned" potable reuse.

Indirect potable water reuse is the abstraction, treatment, and distribution of water for drinking from a natural source water that is fed in part by the discharge of wastewater effluent.

Planned indirect potable water reuse is the purposeful augmentation of a water supply source with reclaimed water derived from treated municipal wastewater. The water receives additional treatment prior to distribution. For example, reclaimed water might be added to ambient water in a water supply reservoir or underground aquifer and the mixture withdrawn for subsequent treatment at a later time.

Unplanned indirect potable reuse is the unintentional addition of wastewater (treated or not) to a water supply that is subsequently used (usually by downstream communities) as a water source, with additional treatment prior to delivery. As noted earlier, many communities already unintentionally practice such unplanned indirect potable reuse.

Direct potable water reuse is the immediate addition of reclaimed wastewater to the water distribution system. This practice has not been adopted by, or approved for, any water system in the United States.

With planned or unplanned indirect potable reuse, the storage provided between treatment and consumption allows time for mixing, dilution, and natural physical, chemical, and biological processes to purify the water. In contrast, with direct potable reuse, the water is reused with no intervening environmental buffer.

With planned indirect potable reuse and direct potable reuse, the wastewater is treated to a much higher degree than it would be were it being discharged directly to a surface water without specific plans for reuse. The wastewater generally is first treated as it would be in a conventional municipal wastewater treatment plant, then subjected to various advanced treatment processes.

Conventional wastewater treatment begins with preliminary screening and grit removal to separate sands, solids, and rags that would settle in channels and interfere with treatment processes (Henry and Heinke, 1989). Primary treatment follows this preliminary screening and usually involves gravity sedimentation. Primary treatment removes slightly more than one-half of the suspended solids and about one-third of the biochemical oxygen demand (BOD) from decomposable organic matter, as well as some nutrients, pathogenic organisms, trace elements, and potentially toxic organic compounds.

Secondary treatment usually involves a biological process. Microorganisms in suspension (in the "activated sludge" process), attached to media (in a "trickling filter" or one of its variations), or in ponds or other processes are used to remove biodegradable organic material. Part of the organic material is oxidized by the microorganisms to produce carbon dioxide and other end products, and the remaining organic material provides the energy and materials needed to support the microorganism community. Secondary treatment processes can remove up to 95 percent of the BOD and suspended solids entering the process, as well as significant amounts of heavy metals and certain organic compounds (Water Pollution Control Federation, 1989). Conventional wastewater treatment usually ends with secondary treatment, except in special cases where tertiary treatment is needed to provide additional removal of contaminants such as microbial pathogens, particulates, or nutrients.

Advanced treatment processes beyond tertiary treatment are necessary when wastewater is to be reclaimed for potable purposes. Table 1-1 provides a list of advanced treatment processes, arranged by the types of constituents they are designed to remove.

The process used by Water Factory 21 in Orange County, California, to treat wastewater prior to injecting it into selected coastal aquifers to form a seawater intrusion barrier is illustrative (see Figure 1-2 and Box 1-2). The advanced treatment of this water includes additional removal of suspended material by chemical coagulation with lime, alum, or a ferric salt. This process is generally quite effective in removing heavy metals as well as dissolved organic materials (McCarty et al., 1980). Recarbonation by the addition of carbon dioxide then neutralizes the high pH created by the addition of lime. After that, mixed media filtration is used to remove suspended solids. The flow is then split between granular activated carbon, which removes soluble organic materials, and reverse osmosis (RO), which is used for demineralization, so that when blended with the remaining water the mixture will meet total dissolved solids requirements specified for injected water. Reverse osmosis can also remove the majority of the dissolved nonvolatile organic materials and achieve less than 1 mg/liter of dissolved organic carbon in the treated water. According to measures of identifiable contaminants, water treated in this manner is often of better quality than some polluted surface waters now used as

TABLE 1-1 Constituent Removal by Advanced Wastewater Treatment Processes

Principal Removal Function	Description of Process	Type of Wastewater Treated[a]
Suspended solids removal	Filtration	EPT, EST
	Microstrainers	EST
Ammonia oxidation	Biological nitrification	EPT, EBT, EST
Nitrogen removal	Biological nitrification/ denitrification	EPT, EST
Nitrate removal	Separate-stage biological denitrification	EPT + nitrification
Biological phosphorus removal	Mainstream phosphorus removal[b]	RW, EPT
	Sidestream phosphorus removal	RAS

TABLE 1-1 Continued

Principal Removal Function	Description of Process	Type of Wastewater Treated[a]
Combined nitrogen and phosphorus removal by biological methods	Biological nitrification/ denitrification and phosphorus removal	RW, EPT
Nitrogen removal by physical or chemical methods	Air stripping	EST
	Breakpoint chlorination	EST + filtration
	Ion exchange	EST + filtration
Phosphorus removal by chemical addition	Chemical precipitation with metal salts or lime	RW, EPT, EBT, EST
Toxic compounds and refractory organics removal	Granular activated carbon adsorption	EST + filtration
	Powdered activated carbon adsorption	EPT
	Chemical oxidation	EST + filtration
Dissolved inorganic solids removal	Chemical precipitation	RW, EPT, EBT, EST
	Ion exchange	EST + filtration
	Ultrafiltration	EST + filtration
	Reverse osmosis	EST + filtration
	Electrodialysis	EST + filtration + carbon adsorption
Volatile organic compounds	Volatilization and gas stripping	RW, EPT
Microorganism removal[c]	Reverse osmosis	EST + filtration
	Nanofiltration/ ultrafiltration	EST + filtration
	Lime treatment	EST

[a]EBT = effluent from biological treatment (before clarification); EPT = effluent from primary treatment; EST = effluent from secondary treatment; RAS = return activated sludge; and RW = raw water (untreated sewage).

[b]Removal process occurs in the main flowstream as opposed to during sidestream treatment.

[c]Microorganism removal is also accomplished by any of several chemical disinfection processes (e.g., free Cl_2, NH_2Cl, ClO_2, O_3), but these are not usually considered as advanced wastewater treatment processes.

SOURCE: Adapted from Metcalf and Eddy, Inc., 1991.

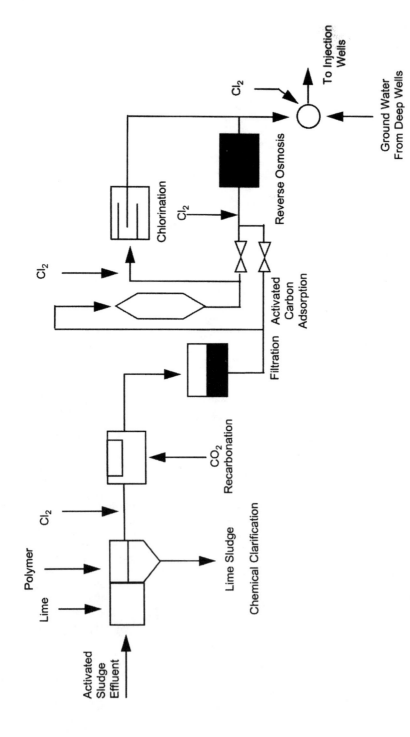

FIGURE 1-2 Flow schematic for Orange County Water District Water Factory 21.

BOX 1-2
Water Factory 21 in Orange County, California

The Orange County Water District (OCWD) has been injecting high quality reclaimed water into selected coastal aquifers to establish a salt water intrusion barrier. Seawater intrusion was first observed in municipal wells during the 1930s as a consequence of basin overdraft. Overdrafting of the ground water continued into the 1950s. Overpumping of the ground water resulted in seawater intrusion as far as 5.6 km (3.5 miles) inland from the Pacific Ocean by the 1960s.

OCWD began pilot studies in 1965 to determine the feasibility of injecting effluent from an advanced wastewater treatment (AWT) facility into potable water supply aquifers. Construction of an AWT facility, known as Water Factory 21, began in 1972 in Fountain Valley, and injection of the treated municipal wastewater into the ground began in 1976.

Water Factory 21 accepts activated-sludge secondary effluent from the adjacent County Sanitation Districts of Orange County wastewater treatment facility. The 5.7×10^7 liter/day (15×10^6 gal/day) water reclamation plant processes consist of lime clarification for removal of suspended solids, heavy metals, and dissolved minerals; recarbonation for pH control; mixed-media filtration for removal of suspended solids; activated carbon absorption for removal of dissolved organic compounds; reverse osmosis for demineralization and removal of other constituents; and chlorination for disinfection and algae control (National Research Council, 1994).

Prior to injection, the product water is blended 2:1 with deep well water from an aquifer not subject to contamination. The blended water is chlorinated in a blending reservoir before it is injected into the ground. Depending on conditions, the injected water flows toward the ocean, forming a seawater barrier; inland to augment the potable ground water supply; or in both directions. On average, well over 50 percent of the injected water flows inland. It is estimated that this injected water makes up no more than 5 percent of the water supply for area residents who rely on ground water.

---

sources of drinking water supply. (The removal of particular chemical and microbiological constituents of concern is discussed in more detail in Chapter 2.)

HISTORY OF PLANNED POTABLE REUSE AND ITS MOTIVATION

Direct potable reuse is not currently approved for use in U.S. water systems. The only documented case of an operational direct potable reuse system is in Windhoek, Namibia, in southern Africa. For 30 years, this facility has been used intermittently to forestall water emergencies during drought conditions (Harhoff and van der Merwe, 1996; see Box 1-3). While direct potable reuse is not practiced in the United States, planned indirect potable reuse is used to augment several U.S. drinking water systems, and pilot facilities have been constructed to evaluate the potential for direct and indirect potable reuse.

The Denver Water Department initiated a series of research projects from 1979 through 1990 on the viability of direct potable reuse of reclaimed water (see Box 1-4). However, Denver presently has no plans for direct potable reuse.

In the Washington, D.C., area, the wastewater-contaminated Potomac River Estuary was evaluated as a potential source of drinking water in an extensive study conducted from 1980 to 1983 by the U.S. Army Corps of Engineers and authorized as part of the Water Resource Development Act of 1974 (see Box 1-1). The Potomac Estuary Experimental Water Treatment Plant was completed in January 1980. The 3.8×10^6 liter/day (1×10^6 gal/day) plant was operated with influent water designed to simulate water quality during drought conditions, when as much as 50 to nearly 100 percent of the estuary flow would consist of treated wastewater. The plant influent contained a blend of Potomac River Estuary water

---

BOX 1-3
Windhoek Direct Water Reclamation System

In 1968, a system for reclaiming potable water from domestic sewage was pioneered in Windhoek, Namibia, to supplement the potable water supply. Surface water sources and ground water extraction had been fully appropriated, and direct reuse of reclaimed water was instituted just in time to avert a water crisis caused by drought. The system has been producing acceptable potable water to the city ever since as part of a larger program to conserve water and manage water demand (Harhoff and van der Merwe, 1996). The reclamation plant has been operated on an intermittent basis to supplement the main supplies during times of peak summer demand or during emergencies.

This system went through a succession of modifications and improvements over the years, accompanied by comprehensive chemical, bacteriological, virological, and epidemiological monitoring. The current sequence of treatment processes involves primary and secondary treatment at the Gammans wastewater treatment plant (primary settling, activated sludge, secondary settling, and maturation ponds). The secondary effluent is then directed to the Goreangab water reclamation plant, where treatment includes alum addition, dissolved air flotation, chlorination, and lime addition, followed by settling, sand filtration, chlorination, carbon filtration, and final chlorination. The final effluent is then blended with treated water from other sources before distribution. Water quality tests are conducted on samples taken from different points in the treatment sequence: in the storage reservoirs, at key points in the distribution system, and at consumer taps. The treated wastewater is also continuously monitored to ensure a consistent, high-quality maturation pond effluent.

BOX 1-4
Denver's Direct Potable Water Reuse
Demonstration Project

In 1968, the Environmental Protection Agency (EPA) allowed Denver to divert water from the Blue River on the west side of the Continental Divide on the condition that it examine a range of alternatives to satisfy projected future demands of a growing metropolitan area. The Direct Potable Water Reuse Demonstration Project was designed to examine the feasibility of converting secondary effluent from a wastewater treatment plant to water of potable quality that could be piped directly into the drinking water distribution system. In 1979, plans were developed for the construction of a demonstration facility to examine the cost and reliability of various treatment processes. The 1.0 mgd (44 liter/s) treatment plant began operation in 1985, and during the first three years, many processes were evaluated (Lauer and Rogers, 1996). Data from the evaluation period were used to select the optimum treatment sequence, which was used to produce samples for a two-year animal feeding health-effect study. Comprehensive analytical studies defined the product water quality in relation to existing standards and to Denver's current potable supply. The product water exceeded the quality of Denver drinking water for all chemical, physical, and microbial parameters tested except for nitrogen, and alternative treatment options were demonstrated for nitrogen removal. The final health-effect study demonstrated no health effects associated with either water. The raw water supply for the reuse plant is unchlorinated secondary effluent (treated biologically) from the metropolitan Denver wastewater treatment facility. Advanced treatment included high-pH lime treatment, single- or two-stage recarbonation, pressure filtration, selective ion exchange for ammonia removal, two-stage activated carbon adsorption, ozonation, reverse osmosis, air stripping, and chlorine dioxide disinfection. Side stream processes included a fluidized bed carbon reactivation furnace, vacuum sludge filtration, and selective ion exchange regenerant recovery.

---

and nitrified secondary effluent from the adjacent Blue Plains Wastewater Treatment Plant.

Artificial recharge is being practiced in some areas of the United States, including recharge of ground water sources with reclaimed water (a form of indirect reuse) in order to replenish depleted underground reserves or prevent salt water intrusion (National Research Council, 1994). Artificial recharge programs began in Whittier Narrows, near Los Angeles, California (see Box 1-5), in the early 1960s using surface percolation of blends of captured storm water, imported water, and treated wastewater in unlined river channels or specially constructed spreading basins (Nellor et al., 1984, 1995). In 1972, Orange County Water District in Fountain Valley, California, began operation of Water Factory 21 (see Box 1-2) to reclaim wastewater for injection into the aquifer as a salt

BOX 1-5
County Sanitation Districts of Los Angeles County
Ground Water Recharge Projects

Since 1962, the Whittier Narrows Water Reclamation Plant (WRP) has used reclaimed water along with surface water and storm water to recharge ground water in the Montebello Forebay area of Los Angeles County by surface spreading of the reclaimed water. The reclaimed water makes up a portion of the potable water supply for the area residents that rely on ground water. From 1962 until 1973, the Whittier Narrows WRP was the sole provider of reclaimed water in the form of disinfected secondary effluent. In 1973, the San Jose Creek WRP began supplying secondary effluent for recharge. Some surplus effluent from a third treatment plant, the Pomona WRP, is released to the San Jose Wash, which ultimately flows to the San Gabriel River and becomes an incidental source for recharge in the Montebello Forebay (Nellor et al., 1984).

The WRPs start their wastewater treatment with primary and secondary biological treatment. In 1978, all three WRPs added tertiary treatment with mono- or dual-media filtration and chlorination/dechlorination to their treatment regimes.

After leaving the reclamation plants, the reclaimed water is conveyed to one of several spreading areas (either specially prepared spreading grounds or dry river channels or washes). In the process of ground water recharge, the water percolates through an unsaturated zone of soil ranging in average thickness from about 3 to 12 m (10 to 40 ft) before reaching the ground-water table. The usual spreading consists of five days of flooding during which water is piped into the basins and maintained at a constant depth. The flow is then discontinued. The basins are then allowed to drain and dry out for 16 days. This wet and dry cycle maintains the proper conditions for the percolation process (Crook et al., 1990; Nellor et al., 1984).

water intrusion barrier. On average, 50 percent of the injected water flows inland to augment the general water supply for Orange County.

In West Texas, the Fred Hervey Water Reclamation Plant began operation in 1985 as a wastewater treatment facility incorporating advanced treatment processes designed for recycling wastewater from the northeast area of El Paso back to the Hueco Bolson aquifer, which supplies both El Paso, Texas, and Juarez, Mexico. This artificial recharge project is necessary to protect the freshwater aquifer from depletion and salt water intrusion. The overall recharge system consists of an advanced wastewater treatment plant, a pipeline system to the injection site, and 10 injection wells to reach the area's deep water table (about 107 m (350 ft) below the surface). After injection, the water travels approximately 1.2 km (0.75 mile) through the aquifer to production wells for municipal water supply (Knorr, 1985).

The city of Phoenix and other municipalities in the Salt River Valley of Arizona are interested in renovating part of their treated municipal wastewater by soil aquifer treatment (SAT) so that it can be stored underground for eventual potable use. The feasibility of SAT in the Phoenix area was studied with a small test project installed in 1967 and a larger demonstration project installed in 1975. The latter could be part of a future operational project that would have a basin area of 48 ha (119 acres) and a projected capacity of about 276×10^6 m^3/year (73×10^9 gal/year). Both projects were operated in the normally dry Salt River bed (Bouwer and Rice, 1984).

Planned augmentation of surface water supplies with reclaimed water is being investigated in both California and the eastern United States for different reasons. San Diego is actively investigating the feasibility of augmenting its general water supplies with reclaimed municipal wastewater because of the high costs of importing water from other parts of the state and the lack of local water sources (see Box 1-6). In Florida, both water shortages and waste disposal requirements are generating increased interest in the use of reclaimed wastewater. Increasingly stringent requirements regulating discharge to sensitive receiving waters have

---

BOX 1-6
San Diego's Total Resource Recovery Project

San Diego, California, imports virtually all of its water supply from other parts of the state. New sources of imported water are not readily available, and the availability of existing supplies is diminishing. The city is thus actively investigating advanced water treatment technologies for reclaiming municipal wastewater that is presently being discharged to the Pacific Ocean. Preliminary experiments were conducted at the bench-scale (0.02×10^6 gal/day) Aqua I facility in Mission Valley from 1981 to 1986. The pilot-scale (0.3×10^6 gal/day secondary, 0.05×10^6 gal/day advanced) treatment Aqua II Total Resource Recovery facility operated at Mission Valley from 1984 through 1992. The full-scale demonstration Aqua III facility (1×10^6 gal/day secondary, 0.5×10^6 gal/day advanced) was constructed in Pasqual Valley and began full-time operation in October 1994.

The Aqua II pilot facility uses channels containing water hyacinths for secondary treatment, followed by a 50,000 gal/day advanced treatment system designed to upgrade the secondary effluent water to a quality equivalent to raw water for potable reuse. The tertiary and advanced process trains were selected in 1985 by a technical advisory committee in conjunction with the city. Tertiary treatment to produce a low-turbidity water suitable for reverse osmosis feedwater was provided by a package water treatment plant, with ferric chloride coagulation, flocculation, sedimentation, and multimedia filtration. The system included ultraviolet light disinfection, cartridge filtration, chemical pretreatment, reverse osmosis using thin-film composite membranes, aeration tower decarbonation, and carbon adsorption. The final process train produces water that meets U.S. drinking water standards.

BOX 1-7
Tampa Water Resource Recovery Project

The Tampa Water Resource Recovery Project was developed to satisfy the future water demands of both the city of Tampa and the West Coast Regional Water Supply Authority. The proposed project involves the supplemental treatment of the Hookers Point Advanced Wastewater Treatment (AWT) Facility effluent to achieve acceptable quality for augmentation of the Hillsborough River raw water supply. A pilot plant was designed, constructed, and operated to evaluate supplemental treatment requirements, performance, reliability, and quality (CH$_2$M Hill, 1993).

Source water for the pilot plant was withdrawn downstream from AWT Facility denitrification filters prior to chlorination. The pilot plant facility evaluated four unit process trains, all of which included preaeration, lime treatment and recarbonation, and gravity filtration, followed by either (1) ozone disinfection, (2) reverse osmosis and ozone disinfection, (3) ultrafiltration and ozone disinfection, or (4) granular activated carbon (GAC) adsorption and ozone disinfection. The process train including GAC adsorption and ozone disinfection was selected for design.

The City of Tampa's industrial base is mostly food oriented. Inputs to the wastewater system were confirmed by a "vulnerability analysis." Tampa has an active pretreatment program, and there has been no interference with the plant's biological process since startup in 1978.

The design of the advanced treatment plant allows for rejection of water at any level of treatment and diversion back to the main plant. The use of a bypass canal for storage and mixing provides a large storage capacity and constant dilution of product water with canal and river water. Water can be diluted from the aquifer when river water is not available. Flood control gates allow the canal to be flushed if a problem is detected. Canal water can be drawn through a "linear well field" along the canal to provide further ground water dilution. Five miles of canal and river provide additional natural treatment prior to the intake for the drinking water treatment plant.

forced many municipal wastewater utilities to upgrade their treatment processes to decrease the level of nutrients in the effluent. This is causing many communities to consider reuse alternatives for municipal wastewater. For example, the City of Tampa has completed a feasibility study (CH$_2$M Hill, 1993) and intends to implement a program to augment its river water supply with reclaimed water (see Box 1-7).

Since 1978, the Upper Occoquan Sewage Authority (UOSA), in northern Virginia, has discharged reclaimed wastewater to the upper reaches of the Occoquan Reservoir, which serves as the principal water supply source for approximately one million people. The UOSA reclamation plant was developed in 1978 in response to deteriorating water quality conditions in the reservoir, which occurred as a result of discharges into the reservoir from several small and poorly operated wastewater treatment plants. The state of Virginia regulates UOSA as a wastewater discharger rather than as a water reclamation facility, though with somewhat more stringent discharge requirements and with recognition of its connection to the water supply. Such indirect reuse may be viewed as similar to the unplanned reuse that occurs when one city discharges its waste into a river or stream used by a downstream community for its water supply.

OVERVIEW OF RELEVANT FEDERAL GUIDELINES AND STATE REGULATIONS

Aside from current drinking water regulations, no enforceable federal regulations specifically address potable reuse. The Environmental Protection Agency (EPA) has developed suggested guidelines for indirect potable reuse (U.S. EPA, 1992), and a few states have developed regulatory criteria. California and Florida are in the forefront of promulgating specific criteria for planned indirect potable reuse. California has prepared draft criteria for ground water recharge, and Florida has adopted criteria for both ground water recharge and surface water augmentation. In Arizona, regulations addressing ground water recharge with treated wastewater are independent from the state's reuse criteria.

Federal Guidelines

EPA's guidance manual on water reuse, though not a formal regulatory document, provides recommendations for a wide range of reuse practices, including indirect potable reuse by ground water recharge or surface water augmentation, that should be useful to state agencies in developing appropriate regulations. Table 1-2 summarizes the suggested criteria related to indirect potable reuse.

In addition to specific wastewater treatment and reclaimed water quality recommendations, the guidelines provide general recommendations to indicate the types of treatment and water quality requirements that are likely to be imposed where indirect potable reuse is contemplated. The guidelines do not include a complete list of suggested water quality limits for all constituents of concern because water quality requirements are constantly changing as new contaminants are added to the list of those regulated under the Safe Drinking Water Act. The guidelines do not advocate direct potable reuse and do not include recommendations for such use.

TABLE 1-2 EPA Suggested Guidelines for Reuse of Municipal
Wastewater

Type of Reuse	Treatment	Reclaimed Water Quality[a]
Ground water recharge by spreading on ground above potable aquifers	• Site-specific • Secondary[c] and disinfection[d] (minimum) • May also need filtration[e] and/or advanced wastewater treatment[f]	• Site-specific • Meet drinking water standards after percolation through vadose zone
Ground water recharge by injection into potable aquifers	• Secondary[c] • Filtration[e] • Disinfection[d] • Advanced wastewater treatment[f]	Includes, but is not limited to, the following: • pH = 6.5-8.5 • Turbidity ≤ 2 NTU[i] • No detectable fecal coliforms per 100 ml[j, k] • Residual[l] ≥ 1 mg/liter Cl_2 • Meet drinking water standards

Reclaimed Water Monitoring	Setback Distances[b]	Comments
Includes, but is not limited to, the following: • pH: daily • Coliform: daily • Cl$_2$ residual: continuous • Drinking water standards: quarterly • Other[g]: depends on constituent	• 600 m (2000 ft) to extraction wells; may vary depending on treatment provided and site-specific conditions	• The depth to ground water (i.e., thickness of the vadose zone) should be at least 2 m (6 ft) at the maximum ground water mounding point • The reclaimed water should be retained underground for at least 1 year prior to withdrawal • Recommended treatment is site specific and depends on factors such as type of soil, percolation rate, thickness of vadose zone, native ground water quality, and dilution • Monitoring wells are necessary to detect the influence of the recharge operation on the ground water • The reclaimed water should not contain measurable levels of pathogens after percolation through the vadose zone[h] • Treatment reliability checks need to be provided
Includes, but is not limited to, the following: • pH: daily • Turbidity: continuous • Coliform: daily • Cl$_2$ residual: continuous • Drinking water standards: quarterly • Other[g]: depends on constituent	• 2000 ft (600 m) to extraction wells; may vary depending on site-specific conditions	• The reclaimed water should be retained underground for at least 1 year prior to withdrawal • Monitoring wells are necessary to detect the influence of the recharge operation on the ground water • Recommended water quality limits should be met at the point of injection • The reclaimed water should not contain measurable levels of pathogens at the point of injection[h]

Table continues on next page

TABLE 1-2 Continued

Type of Reuse	Treatment	Reclaimed Water Quality[a]
Augmentation of surface supplies	• Secondary[c] • Filtration[d] • Disinfection[e] • Advanced wastewater treatment[f]	Includes, but is not limited to, the following: • pH = 6.5-8.5 • Turbidity ≥ 2 NTU[i] • No detectable fecal coliforms per 100 ml[j,k] • Residual[l] ≥ 1 mg/liter Cl_2 • Meet drinking water standards

NOTE: NTU = nephelometric turbidity units.

[a]Unless otherwise noted, recommended quality limits apply to reclaimed water at the point of discharge from the treatment facility.

[b]Setbacks are recommended to protect potable water supply sources from contamination and to protect humans from unreasonable health risks due to exposure to reclaimed water.

[c]Secondary treatment processes include activated sludge, trickling filters, rotating biological contactors, and many stabilization pond systems. Secondary treatment should produce effluent in which both the BOD and suspended solids do not exceed 30 mg/liter.

[d]Disinfection means the destruction, inactivation, or removal of pathogenic microorganisms by chemical, physical, or biological means. Disinfection may be accomplished by chlorination, ozonation, other chemical disinfectants, ultraviolet radiation, membrane processes, or other processes.

[e]Filtration means the passing of wastewater through natural undisturbed soils or filter media such as sand and/or anthracite.

[f]Advanced wastewater treatment processes include chemical clarification, carbon adsorption, reverse osmosis and other membrane processes, air stripping, ultrafiltration, and ion exchange.

Reclaimed Water Monitoring	Setback Distances[b]	Comments
		• A higher chlorine residual and/or a longer contact time may be necessary to ensure virus inactivation • Treatment reliability checks need to be provided
Includes, but is not limited to, the following: • pH: daily • Turbidity: continuous • Coliform: daily • Cl$_2$ residual: continuous • Drinking water standards: quarterly • Other[g]: depends on constituent	• Site-specific	• Recommended level of treatment is site-specific and depends on factors such as receiving water quality, time and distance to point of withdrawal, dilution, and subsequent treatment prior to distribution for potable uses • The reclaimed water should not contain measurable levels of pathogens[h] • A higher chlorine residual and/or a longer contact time may be necessary to ensure virus inactivation • Treatment reliability checks need to be provided

[g]Monitoring should include measurement of the concentrations of inorganic and organic compounds, or classes of compounds, that are known or suspected to be toxic, carcinogenic, teratogenic, or mutagenic and are not included in the drinking water standards.

[h]It is advisable to fully characterize the microbiological quality of the reclaimed water prior to implementation of a reuse program.

[i]The recommended turbidity limit should be met prior to disinfection. The average turbidity should be based on a 24-hour time period. The turbidity should not exceed 5 NTU at any time. If suspended solids content is used in lieu of turbidity, the average suspended solids concentration should not exceed 5 mg/liter.

[j]Unless otherwise noted, recommended coliform limits are median values determined from the bacteriological results of the last seven days for which analyses have been completed. Either the membrane filter or the fermentation tube technique may be used.

[k]The number of fecal coliform organisms should not exceed 14/100 ml in any sample.

[l]Total chlorine residual after a minimum contact time of 30 minutes.

SOURCE: Adapted from U.S. EPA, 1992.

California Wastewater Reclamation Criteria

California currently includes general requirements for indirect potable water reuse via ground water recharge under the state's Wastewater Reclamation Criteria (State of California, 1978). These requirements are presently being replaced with more detailed regulations focusing specifically on ground water recharge (State of California, 1993). The proposed regulations, which have gone through several iterations, are designed to ensure that ground water extracted from an aquifer recharged by reclaimed water meets all drinking water standards and requires no treatment prior to distribution. Table 1-3 summarizes the proposed treatment process and site requirements. The criteria are intended to apply to any water reclamation project designed for the purpose of recharging ground water suitable for use as a drinking water source (Hultquist, 1995).

The proposed regulations prescribe both microbiological and chemical constituent limits, some of which are summarized in Table 1-3. The proposed regulations would require that concentrations of minerals, trace inorganic chemicals, and organic chemicals in reclaimed water prior to recharge must not exceed the maximum contaminant levels established in the state's drinking water regulations. The total nitrogen concentration of the reclaimed water cannot exceed 10 mg/liter unless it is demonstrated that in the process of percolating into the ground water, enough nitrogen will be removed from the reclaimed water to meet the 10 mg/liter standard.

Based principally on information and recommendations contained in a report prepared by an expert panel commissioned by California (State of California, 1987), the proposed regulations specify that extracted ground water should contain no more than 1 mg/liter of total organic carbon (TOC) of wastewater origin. TOC is considered to be a suitable measure of the gross organics content of reclaimed water for the purpose of determining organics removal efficiency in practice. The requirements shown in Table 1-3 are intended in part to ensure that the TOC concentration of wastewater origin is limited to 1 mg/liter in public water supply wells. Requirements for reduction of TOC concentrations are less restrictive for projects in which the reclaimed water is recharged into the ground via surface spreading than for projects in which the reclaimed water is injected directly into the aquifer, because additional TOC removal has been demonstrated to occur in the unsaturated zone with surface spreading projects (Nellor et al., 1984). Similarly, the proposed regulations require that the composition of the water at the point of extraction not exceed either 20 percent or 50 percent water of reclaimed water origin, depending on site-specific conditions, type of recharge, and treat-

TABLE 1-3 Proposed California Ground Water Recharge Criteria

Treatment and Recharge Site Requirements	Project Category[a]			
	I	II	III	IV
Required treatment				
Secondary	X[b]	X	X	X
Filtration	X	X		X
Disinfection	X	X	X	X
Organics removal	X			X
Maximum allowable reclaimed water in extracted well water (%)	50	20	20	20
Depth to ground water at initial percolation rate of				
<0.5 cm/min (<0.2 in/min)	3 m (10 ft)	3 m (10 ft)	6 m (20 ft)	n.a.[c]
<0.8 cm/min (<0.3 in/min)	6 m (20 ft)	6 m (20 ft)	15 m (50 ft)	n.a.[c]
Minimum retention time underground (months)	6	6	12	12
Horizontal separation[d]	150 m (500 ft)	150 m (500 ft)	300 m (1000 ft)	600 m (2000 ft)

[a]Categories I, II, and III are for surface spreading projects with different levels of treatment. Category IV is for injection projects.
[b]X means that the treatment process is required.
[c]Not applicable.
[d]From edge of recharge operation to the nearest potable water supply well.

SOURCE: Adapted from State of California, 1993.

ment provided. The proposed dilution requirement must be met at all extraction wells.

To ensure removal of pathogens and trace organic constituents in surface spreading operations, the criteria include standards regarding percolation rates and depth to ground water. These standards are intended to provide unsaturated vadose zones that will allow the development of aerobic biological processes that retain or degrade organic chemicals and remove microorganisms from the water. The proposed minimum vadose zone depth varies from 3 m (10 ft) to 15 m (50 ft) depending on site-specific conditions and treatment. Studies have shown that a soil's initial percolation capacity must be less than 0.8 cm/min (0.3 in/min) if it is to provide these benefits (State of California, 1979). If a soil's initial percolation capacity is less than 0.5 cm/min (0.2 in/min), the criteria provide an additional "credit" for soil column treatment that reduces the required vadose zone percolation distance. Maximum percolation capacities are to be determined from initial percolation test results conducted before the recharge operation starts and not from equilibrium infiltration rates (Hultquist, 1995).

The proposed criteria for minimum underground retention time are designed to ensure further die-off or removal of enteric viruses. The retention times are typical of those in current projects judged by state regulators to be safe (Hultquist, 1995). The criteria call for the actual retention time underground to be determined annually at the first (in time) domestic water supply well to receive reclaimed water. The California Department of Health Services does not quantify the expected level of virus reduction underground. Rather, the retention time requirement simply provides an extra barrier to virus survival.

California has not developed criteria for indirect potable reuse via surface water augmentation, although a framework has been proposed (California Potable Reuse Committee, 1996). Augmentation of surface drinking water sources with reclaimed water in California requires two state permits—a waste discharge or reclamation permit from a California Regional Water Quality Control Board and an amended water supply permit from the Department of Health Services.

Florida Water Reuse Requirements

Until the late 1970s, the primary force driving implementation of reuse projects in Florida was effluent disposal. The state's first reuse-related regulations addressed the land application of municipal wastewater (Florida Department of Environmental Regulation, 1983). In the late 1970s, however, demand for water supplies increased, treated wastewater began to be viewed as a drinking water resource, and the state embarked on a program to encourage water reuse and develop regulations that would provide appropriate public health and environmental protection. In 1989, Florida added a chapter entitled "Reuse of Reclaimed Water and Land Application" to its administrative code; these regulations have since been revised (Florida Department of Environmental Protection, 1996). Surface water augmentation is covered by Chapter 62-610 of the Florida Administrative Code (F.A.C.), entitled "Reclaimed Water and Land Application," and Chapter 62-600 F.A.C., entitled "Domestic Wastewater Facilities." Florida now requires the state's water management districts to identify water resource "caution areas" that have critical water supply problems or that are anticipated to have critical problems within the next 20 years (Florida Department of Environmental Protection, 1995). State legislation requires preparation of water reuse feasibility studies for wastewater treatment facilities located within such caution areas and requires a "reasonable" amount of reclaimed water use unless such reuse is not economically, environmentally, or technically feasible. In addition, if reuse is found to be feasible, disposal by surface water discharge or deep well injection is limited to backups for reuse systems.

Table 1-4 summarizes Florida's requirements for reclaimed water used to augment potable water sources. Daily monitoring is required for fecal coliform organisms, carbonaceous biochemical oxygen demand (CBOD), and total suspended solids (TSS). The allowable limits for coliforms, CBOD, and TSS, as well as treatment requirements, vary depending on how the reclaimed water is discharged into the water supply source and the characteristics of the water source.

The first types of water reuse shown in Table 1-4, rapid-rate infiltration basin systems and absorption field systems, have less stringent water quality limits and treatment requirements than do the other types of reuse because the water receives some treatment as it percolates through the soil. Any wastewater land application system located over a potential source of drinking water must meet these standards. For absorption fields, a more stringent TSS limitation of 10 mg/liter may be imposed to protect against formation plugging. Loading to these systems is limited to 23 cm/day (9 in/day), and wetting and drying cycles must be used. For systems having higher loading rates or unfavorable geologic conditions that rapidly move reclaimed water into aquifers, the reclaimed water must receive secondary treatment, filtration, and high-level disinfection and must meet primary and secondary drinking water standards. These criteria are similar to those in the California regulations for surface spreading of reclaimed water.

The other types of water reuse shown in Table 1-4 involve rapid infiltration of reclaimed water into basins in which soil percolation will not provide appreciable additional treatment, direct injection into ground water, and discharge to class I surface waters used for potable supply. Accordingly, such waters must meet stricter standards regarding detectable fecal coliforms, total suspended solids, and chlorine residuals. The rules acknowledge that higher chlorine residuals and/or longer contact times may be needed to meet the fecal coliform requirement.

For augmentation of surface water sources, outfalls for discharge of reclaimed water cannot be located within 150 m (500 ft) of a potable water intake.

Water quality and treatment requirements are most stringent for injection into formations of the Floridian and Biscayne aquifers where total dissolved solids (TDS) do not exceed 500 mg/liter. For these situations

TABLE 1-4 Florida Treatment and Quality Criteria for Reclaimed
Water

Type of Use	Water Quality Limits	Treatment Required
Rapid infiltration basins and absorption fields	200 fecal coliform/100 ml 20 mg/liter TSS[a] 20 mg/liter CBOD[b] 12 mg/liter NO_3 (as N)	Secondary plus disinfection
Rapid infiltration basins in unfavorable geohydrologic conditions	No detectable fecal coliforms/100 ml[c] 5.0 mg/liter TSS Primary and secondary U.S. drinking water standards	Secondary, filtration, and disinfection
Injection to ground water	No detectable fecal coliforms/100 ml[a] 5.0 mg/liter TSS Primary and secondary U.S. drinking water standards	Secondary, filtration, and disinfection
Injection to formations of Floridian or Biscayne aquifers having TDS <500 mg/liter	No detectable fecal coliforms/100 ml[a] 5.0 mg/liter TSS 5 mg/liter TOC 0.2 mg/liter TOX[d] Primary and secondary U.S. drinking water standards	Secondary, filtration, disinfection, and activated carbon adsorption
Discharge to class I surface waters used for potable supply	No detectable fecal coliforms/100 ml[a] 5 mg/liter TSS 20 mg/liter CBOD 10 mg/liter NO_3 (as N) Primary and secondary U.S. drinking water standards	Secondary, filtration, and disinfection

[a]TSS = total suspended solids.
[b]CBOD = carbonaceous BOD.
[c]No detectable fecal coliform organisms per 100 ml in at least 75% of the samples, with no single sample to exceed 25 fecal coliform organisms/100 ml.
[d] TOX = total organic halogen.

SOURCE: Florida Department of Environmental Protection, 1993, 1996.

the regulations specify that reclaimed water must meet drinking water standards, be treated with activated carbon adsorption to remove organics, and have average TOC and total organic halogen (TOX) concentrations less than 5.0 mg/liter and 0.2 mg/liter, respectively. The rules also require that such systems undergo two years of full-scale operational testing.

The Florida Department of Environmental Protection (DEP) is currently refining the requirements for indirect potable reuse. The DEP is considering allowing streamlined pilot testing requirements for projects involving injection into formations of the Floridian and Biscayne aquifers where the TDS does not exceed 500 mg/liter. In addition, the average and maximum TOC limits may be reduced to 3 mg/liter and 5 mg/liter, respectively. Strict limits on TOC and TOX that are currently applicable only to high-quality (TDS < 500 mg/liter) portions of the Floridian and Biscayne aquifers may be extended to a wider range of injection applications.

Arizona Water Reuse Regulations

Arizona's water reclamation and reuse regulations specifically prohibit the use of reclaimed water for direct human consumption (State of Arizona, 1991). Ground water recharge projects are regulated by the Arizona Department of Environmental Quality (ADEQ) and the Arizona Department of Water Resources (ADWR).

In general, ADEQ regulates ground water quality and ADWR manages ground water supply. These agencies require several different permits for any ground water recharge project. A ground water recharge project must obtain an aquifer protection permit from ADEQ. Additionally, both the owner of the wastewater treatment plant that provides the reclaimed water for ground water recharge and the owner or operator of the ground water recharge project that uses the reclaimed water must obtain permits from the ADWR before any reclaimed water can be recharged (Arizona Department of Water Resources, 1995). A single permit may be issued if the same applicant applies for both permits and the permits are sought for facilities located in a contiguous geographic area.

To obtain an aquifer protection permit from ADEQ, the recharge project applicant must demonstrate that the project will not cause or contribute to a violation of an aquifer water quality standard. If aquifer water quality standards are already being violated in the receiving aquifer, the permit applicant must demonstrate that the ground water recharge project will not further degrade aquifer water quality. All aquifers in Arizona currently are classified for drinking water use, and the state has adopted National Primary Drinking Water Maximum Contaminant Levels (MCLs) as aquifer water quality standards. These standards apply to all ground water in saturated formations yielding more than 20 liters/day (5 gal/day) of water (which is essentially all ground water in Arizona). Thus, reclaimed water must be treated to meet drinking water standards before it can be injected into an aquifer.

A ground water recharge project that uses reclaimed water is also required to obtain an underground storage facility permit from ADWR. To get this permit the applicant must demonstrate that (1) the applicant possesses the technical and financial capability to construct and operate the ground water recharge project; (2) the aquifer contains sufficient capacity for the maximum amount of reclaimed water that could be in storage at any one time; (3) the storage of reclaimed water will not cause unreasonable harm to land or to other water users; (4) the applicant has applied for and received any required floodplain use permit from the county flood control district; and (5) the applicant has applied for and received an aquifer protection permit from ADEQ. If received, the underground storage facility permit will prescribe the design capacity of the ground water recharge project, the maximum annual amount of reclaimed water that may be stored, and monitoring requirements.

Before recovering any of the reclaimed water that has been stored underground, the person or entity seeking to recover the water must apply to ADWR for a recovery well permit. If the recovery well permit is for a new well, ADWR must determine that the proposed recovery of the stored water will not unreasonably increase damage to surrounding land or other water users. If the recovery well permit is for an existing well, the applicant must demonstrate that it has a right to use the existing well. A recovery well permit includes provisions that specify the maximum pumping capacity of the recovery well.

CONCLUSIONS

The historical approach to water supply development has been to withdraw water from the best available source. In some parts of the United States, however, high-quality source waters are becoming increasingly scarce, and some municipalities are using or are beginning to consider using reclaimed municipal wastewater to augment their potable water supplies. While the maxim that drinking water should be obtained from the best available source should still be the guiding principle for water supply development, in some instances the best available source of additional water to augment natural sources of supply may be reclaimed water. No enforceable federal regulations currently govern the use of reclaimed water for potable purposes, and only a few states have developed detailed criteria for water reuse. Any water utility considering a potable water reuse project should carefully consider the public health, water treatment, and quality assurance issues discussed in this report to ensure that its consumers are protected from any potential adverse effects of water reuse.

REFERENCES

Arizona Department of Water Resources. 1995. Environmental Quality Act. Arizona Revised Statutes Section 49-241. Phoenix: Arizona Department of Water Resources.

Bouwer, H., and R. C. Rice. 1984. Renovation of wastewater at the 23rd Avenue Rapid Infiltration Project. Journal-Water Pollution Control Federation 56(1):76-83.

California Potable Reuse Committee. 1996. A proposed framework for regulating the indirect potable reuse of advance treated reclaimed water by surface water augmentation in California. Sacramento: California Department of Water Resources.

CH_2M Hill. 1993. Tampa Water Resource Recovery Project Pilot Studies. Tampa, Fla.: CH_2M Hill.

Crook, J., T. Asano, and M. H. Nellor. 1990. Groundwater recharge with reclaimed water in California. Water Environment and Technology 2(8):42-49.

Ding, W., Y. Fujita, E. Aeschimann, and M. Reinhard. 1996. Identification of organic residues in tertiary effluents by GC/EI-MS, GS/CI-MS, and GC/TSQ-MS. Fresenius Journal of Analytical Chemistry 354:48-55.

Florida Department of Environmental Protection. 1993. Domestic wastewater facilities. Chapter 62-600, Florida Administrative Code. Tallahassee: Florida Department of Environmental Protection.

Florida Department of Environmental Protection. 1995. State water policy. Chapter 62-40, Florida Administrative Code. Tallahassee: Florida Department of Environmental Protection.

Florida Department of Environmental Protection. 1996. Reuse of Reclaimed Water and Land Application. Chapter 62-610, Florida Administrative Code. Tallahassee: Florida Department of Environmental Protection.

Florida Department of Environmental Regulation. 1983. Land Application of Domestic Wastewater. Tallahassee: Florida Department of Environmental Regulation.

Harhoff, J., and B. van de Merwe. 1996. Twenty-five years of wastewater reclamation in Windhoek, Namibia. Water Science and Technology 33(10-11):25-35.

Henry, J. G., and G. W. Heinke. 1989. Environmental Science Engineering. New York: Prentice Hall.

Hultquist, R. H. 1995. Augmentation of ground and surface drinking water sources with reclaimed water in California. Paper presented at AWWA Annual Conference, Workshop on Augmenting Potable Water Supplies with Reclaimed Water. June 18, 1995, Fountain Valley, Calif.

Knorr, D. B. 1985. Status of El Paso, Texas recharge project. Pp. 137-152 In Proceedings of Water Reuse Symposium III. Denver, Colo.: American Water Works Association.

Lauer, W.C., and S. E. Rogers. 1996. The demonstration of direct potable reuse: Denver's pioneer project. Pp. 269-289 in AWWA/WEF 1996 Water Reuse Conference Proceedings. Denver: American Water Works Association.

McCarty, P. L., M. Reinhard, J. Graydon, J. Schreiner, K. Sutherland, T. Everhart, and D. G. Argo. 1980. Wastewater Contaminant Removal for Groundwater Recharge. EPA-600/2-80-114. Cincinnati, Oh.: U.S. Environmental Protection Agency.

Metcalf and Eddy, Inc. 1991. Wastewater Engineering: Treatment, Disposal, and Reuse. 3rd Ed. New York: McGraw-Hill.

National Research Council. 1982. Quality Criteria for Reuse. Washington, D.C.: National Academy Press.

National Research Council. 1984. The Potomac Estuary Experimental Water Treatment Plant. Washington, D.C.: National Academy Press.

National Research Council. 1994. Ground Water Recharge Using Waters of Impaired Quality. Washington, D.C.: National Academy Press.

Nellor, M. H., R. B. Baird, and J. R. Smyth. 1984. Health Effects Study—Final Report. Whittier, Calif.: County Sanitation Districts of Los Angeles County.

Nellor, M. H., R. B. Baird, and J. R. Smyth. 1995. Health effects of indirect potable water reuse. Journal of the American Water Works Association 77(7):88-89.

State of Arizona. 1991. Regulations for the Reuse of Wastewater. Arizona Administrative Code, Chapter 9, Article 7. Phoenix: Arizona Department of Environmental Quality.

State of California. 1975. A "state-of-the-art" review of health aspects of wastewater reclamation for ground water recharge. Report prepared by the State of California Water Resources Control Board, Department of Water Resources, and Department of Health. Published by the State of California Department of Water Resources, Sacramento, Calif.

State of California. 1978. Wastewater Reclamation Criteria. California Administrative Code, Title 22, Division 4, California Department of Health Services, Sanitary Engineering Section, Berkeley, Calif.

State of California. 1979. Minimum Guidelines for the Control of Individual Wastewater Treatment and Disposal Systems. California Regional Water Quality Control Board, San Francisco Bay Region, Oakland, Calif.

State of California. 1987. Report of the Scientific Advisory Panel on Ground Water Recharge with Reclaimed Water. G. Robeck (ed.). Prepared for the State of California Water Resources Control Board, Department of Water Resources, and Department of Health Services. Published by the State of California Department of Water Resources, Sacramento, Calif.

State of California. 1993. Draft Proposed Groundwater Recharge Regulation. Prepared by the State of California Department of Health Services, Division of Drinking Water and Environmental Management, Sacramento, Calif.

Swayne, M., G. Boone, D. Bauer, and J. Lee. 1980. Wastewater in Receiving Waters at Water Supply Abstraction Points. EPA-60012-80-044. Washington, D.C.: U.S. Environmental Protection Agency.

U.S. Environmental Protection Agency. 1975. National Interim Primary Drinking Water Regulations. Fed. Reg. 40(248), 59566-59587 (Dec. 24, 1975).

U.S. Environmental Protection Agency. 1992. Guidelines for Water Reuse. EPA/625/R-92/004, U.S. Environmental Protection Agency, Center for Environmental Research Information, Cincinnati, Oh.

Water Pollution Control Federation. 1989. Water Reuse: Manual of Practice, 2nd ed. Alexandria, Va.: Water Environmental Federation.

2

Chemical Contaminants in Reuse Systems

Municipal wastewater contains many chemicals that present known or potential health risks if ingested. The concentration of these contaminants must be reduced before such water is used to augment a water supply.

The mix of chemicals in wastewater varies depending on what types of industries and land uses the service area includes, the nature of the wastewater collection system, and the effectiveness of industrial pretreatment and source control programs. As summarized in Table 2-1, wastewaters contain known inorganic chemicals and minerals that are present naturally in the potable water supply; chemicals from industrial, commercial, and other human activities in the wastewater service area; and chemicals added or generated during water and wastewater treatment and distribution. In addition, unidentified or poorly characterized synthetic organic compounds, derivatives, and breakdown products may be present at potentially harmful levels.

Any of the chemical types shown in Table 2-1 might pose some long-term risk, and the risks may change from one location and time to the next. The ability to evaluate and manage those risks is greatest for minerals and trace inorganic chemicals, less for identifiable organic compounds and disinfection by-products, and minimal for the unidentified mix that comprises the majority of the organics in the water.

45

TABLE 2-1 A Categorization of Chemical Constituents in
Wastewater

Category	Examples
Recognized Chemical Constituents	
Naturally occurring minerals and inorganic chemicals, generally at concentrations greater than 1 mg/liter	Chloride, sodium, sulfate, magnesium, calcium, phosphorus, nitrogen
Chemicals of anthropogenic origin, generally at concentrations less than 1 mg/liter	Regulated contaminants and priority pollutants (trace inorganic and organic chemicals)
Chemicals generated as a result of water and wastewater treatment	Known disinfection by-products, humic substances
Unknown or of Potential Concern	
Possibly present as a component of organic mixtures	Proprietary chemicals and mixtures from industrial applications and their metabolites; unidentified halogenated compounds (unknown disinfection by-products); pharmaceuticals; endocrine disruptors

RECOGNIZED CHEMICAL CONTAMINANTS

Naturally occurring minerals, such as calcium, sulfate, and magnesium, are typically present in most conventional water supplies at concentrations greater than 1 mg/liter. They are regulated according to secondary U.S. drinking water standards, based primarily on their aesthetic effects. The concentrations of these chemical species and of nitrogen-containing compounds and other inorganic salts increase as the water is used and then collected as wastewater. However, the potential hazards they pose to downstream consumers remain manageable because these minerals and salts can be accurately quantified in water, and well-established treatment processes can usually reduce their concentrations to levels complying with national drinking water standards or recommended limits.

Levels of phosphate and nitrogen, two other chemicals commonly found in wastewater, are often monitored at treatment plants because of their potential effects on the ecology of receiving waters. Phosphorus can be efficiently removed from wastewater by chemical precipitation or various biological processes, and nitrogen can be removed by biological nitrification and denitrification or by ion exchange either before or after nitrification.

Although the comparable data regarding trace inorganics (e.g., metals) and specific identifiable organic contaminants (including some disinfection by-products) are less extensive than those for major inorganic ions such as minerals and salts, substantial research and practical experience do exist regarding these compounds in municipal wastewater and their removal during waste treatment. For instance, removals of some priority pollutants and other potentially toxic organic compounds in wastewater treatment plants have been reported by a number of researchers, including Richards and Shieh (1986), Hannah et al. (1986), and Petrasek et al. (1982).

The ability of advanced wastewater treatment (AWT) processes to remove many trace chemical contaminants is well established. Numerous potable reuse studies have shown that AWT can produce water that meets U.S. drinking water standards. Table 2-2, for example, compares the quality of water produced by San Diego's Aqua III pilot plant, Tampa's Hookers Point AWT pilot plant, and Denver's Potable Reuse Demonstration Project to drinking water standards. (See Chapter 1, Boxes 1-4, 1-6, and 1-7 for a description of the treatment processes used in those AWT facilities.)

Most of the potable reuse projects reviewed in this report have conducted extensive analyses for identifiable organic compounds, including

The Aqua II pilot facility, used to demonstrate the feasibility of wastewater reclamation for San Diego, California. Photo courtesy of the City of San Diego.

TABLE 2-2 Comparison of Inorganic and General Water Quality
Parameters of Three Reclaimed Water Systems and U.S. Drinking
Water Standards

Constituent	U.S. Drinking Water Standards	Reclaimed Water		
		San Diego	Tampa	Denver
Physical				
TOC	—	0.27	1.88	0.2
TDS	500	42	461	18
Turbidity (NTU)	—	0.27	0.05	0.06
Nutrients				
Ammonia-N	—	0.8	0.03	5
Nitrate-N	—	0.6	0	0.1
Phosphate-P	—	0.1	0	0.02
Sulfate	250	0.1	0	1
Chloride	250	15	0	19
TKN	—	0.9	0.34	5
Metals				
Arsenic	0.05	<0.0005	0[a]	ND[b]
Cadmium	0.005	<0.0002	0[a]	ND
Chromium	0.1	<0.001	0[a]	ND
Copper	1.0	0.011	0[a]	0.009
Lead	—[c]	0.007	0[a]	ND
Manganese	0.05	0.008	0[a]	ND
Mercury	0.002	<0.0002	0[a]	ND
Nickel	0.1	0.0007	0.005	ND
Selenium	0.05	<0.001	0[a]	ND
Silver	0.05	<0.001	0[a]	ND
Zinc	5.0	0.0023	0.008	0.006
Boron	—	0.29	0	0.2
Calcium	—	<2.0	—	1.0
Iron	0.3[d]	0.37	0.028	0.02
Magnesium	—	<3.0	0	0.1
Sodium	—	11.9	126	4.8

NOTES: NTU = nephelometric turbidity units; TDS = total dissolved solids; TKN = total
Kjeldahl nitrogen; TOC = total organic carbon. San Diego physical and nutrient concentration values are arithmetic means. Any nondetected observations were assumed to be
present at the corresponding detection limit. Metal concentration values are geometric
means determined through probit analysis. Tampa values are arithmetic means of detected
values. Denver values are geometric means of detected values.

[a]Not detected in seven samples.
[b]Not detected in more than 50% of samples.
[c]Lead is regulated according to a treatment standard.
[d]Noncorrosive limit for iron.

SOURCE: CH_2M Hill, 1993; Lauer et al., 1991; Western Consortium for Public Health,
1997.

priority organic pollutants regulated under U.S. drinking water standards as well as additional compounds of concern. The organic analytes evaluated by San Diego included 62 volatile organic compounds; 68 semivolatile organic compounds, including trihalomethanes, benzene, N-nitrosamines, chlorinated aromatics, phenols, and polynuclear aromatic hydrocarbons; pesticides and polychlorinated biphenyls (PCBs); chlorinated dibenzodioxins/dibenzofurans; and low molecular weight aldehydes (Western Consortium for Public Health, 1997). Concentrations of all regulated contaminants were below U.S. and state drinking water standards. Similar evaluations of organic chemicals, with similar results, were conducted at Tampa (CH$_2$M Hill, 1993) and Denver (Lauer et al., 1991).

The Denver reuse project conducted an organic challenge study in which 15 different organic compounds were dosed at approximately 100 times the normal levels found in the reuse plant influent (Lauer et al., 1991). Table 2-3 shows the initial doses and removal rates of these compounds for four different treatment processes. Five of the compounds were removed completely (i.e., to below detectable limits) by lime treatment, and eight of the remaining ten were removed completely by the granular activated-carbon filters. The reverse-osmosis membranes allowed 1.1 mg/liter of chloroform to pass through; this chloroform was subsequently removed by air stripping. The study showed that even

TABLE 2-3 Reuse Plant Organic Challenge Study (cumulative % removals)

Compound	Initial Dose (mg/liter)	Lime	Carbon	Reverse Osmosis	Plant Effluent
Acetic acid	5054	100	—	—	—
Anisole	23	100	—	—	—
Benzothiazole	86.2	63	100	—	—
Chloroform	229.6	26	99.7	99.9	100
Clofibric acid	17.1	0	100	—	—
Ethyl benzene	25.1	100	—	—	—
Ethyl cinnamate	67.8	100	—	—	—
Methoxychlor	44.6	84	100	—	—
Methylene chloride	230	8	100	—	—
Tributyl phosphate	69.4	51	100	—	—
Toluene[a]		25	97	100	—
Benzene[a]		40	100	—	—
Ethylbenzene[a]		36	100	—	—
Xylene[a]		32	100	—	—

[a]Dosed as 2115 mg/liter of gasoline.

SOURCE: Modified from Lauer et al., 1991.

when the given organic compounds are dosed at 100 times the normal concentrations, the AWT processes can remove the contaminants to nondetectable levels.

Orange County Water District conducts an extensive monitoring program to measure organics at various steps within Water Factory 21's treatment process. In more than 18 years of research, only 25 of the 100 organic priority pollutants analyzed have been routinely present above detection limits in the secondary effluent feedwater. The organics detected typically include disinfection by-products, such as trihalomethanes (THMs), at concentrations substantially lower than the current federal drinking water maximum contaminant level (MCL) of 100 µg/liter of total THMs. THMs have not been detected in the monitoring wells. Levels of other priority pollutants in the final product water were less than 1 µg/liter in 1995 (Mills, 1996).

Los Angeles County has conducted long-term analysis of organic contaminants at the Montebello Forebay, where artificial recharge with reclaimed water has been conducted since 1962. The most recent reports on this project show that the average concentration of target organic compounds has not exceeded the most stringent water quality standards and guidelines (Water Replenishment District of Southern California, 1996).

Effective source control programs, enforcement and monitoring of water quality standards, and reliability of water and wastewater treatment systems are standard measures for the protection of public health. Although a water reclamation plant could fail, monitoring at the water treatment plant would probably identify elevated levels of a regulated contaminant if such events occurred with some frequency. For most of these contaminants, the risk is associated with lifetime contamination rather than acute toxicity at the low levels likely to be present. Given the low probability that a spike of a regulated contaminant would pass unnoticed through the wastewater treatment plant, the environmental buffer, and the water treatment plant, along with the small likelihood that it would have acute effects on consumers, the risks associated with this type of event are small.

As a result, identifiable and quantifiable contaminants in wastewater (which include inorganic contaminants, radionuclides, organic priority pollutants, and many other trace organic compounds) pose a manageable risk with respect to their appearance in finished potable water.

MANAGING CHEMICAL INPUTS TO REUSE SYSTEMS

One of the most effective ways to manage and reduce chemical contamination of drinking water systems is effective source control of pollutants by pollution prevention and industrial pretreatment programs. In this regard, the health concerns for planned potable reuse projects differ little from those associated with unplanned potable reuse, where drinking water is obtained from sources that receive upstream wastewater discharge. The 1996 amendments to the Safe Drinking Water Act make source water protection a national priority by encouraging a prevention orientation to pollution control. The law creates a new program and funding for states to conduct assessments of source water areas to determine the susceptibility of drinking water sources to contamination. The Environmental Protection Agency (EPA) is currently engaging stakeholders to develop guidelines for implementing the program.

The industrial pretreatment provisions of the Clean Water Act and its 1990 amendments (42 U.S.C. 13101, et seq.) established maximum allowable concentrations that limit discharges of priority pollutants to wastewater treatment plants. Municipalities have authority to limit the discharges of other potentially harmful compounds on a case-by-case basis. The Clean Water Act also limits the discharge of these compounds from treatment plants (and other direct dischargers) to receiving waters. The existence of such permitting programs allows regulators to identify pollutants of concern and to enforce risk reduction strategies. The Clean Water Act also requires users of the chemicals to improve their management practices and/or to develop effective treatment processes.

For example, the County Sanitation Districts of Los Angeles County (CSDLAC) take steps to exclude from the sewer system certain contaminants that might adversely impact the quality of the reclaimed water being produced. The primary measure aimed at protecting water quality is CSDLAC's industrial waste pretreatment program, created in 1972. As illustrated in Table 2-4, these measures have significantly reduced loadings to the project's wastewater treatment plant. The program presently regulates an extensive and varied industrial base consisting of over 3400 industrial users. It controls noncompatible waste discharges through rigorous up-front permitting and pretreatment requirements, field presence by the inspection staff and monitoring crews, and aggressive enforcement actions for all violations. To further protect effluent quality, industrial discharges have been diverted to "nonreclaimable waste lines" wherever possible. These interceptor sewers typically divert predominately residential wastewater to the water reclamation plants or route industrial wastewater around the reclamation plants to the wastewater treatment plant for ocean disposal. Thus, the water reclamation plants treat mainly residential and commercial waste, with less than 10 percent of the influent coming from industrial sources.

As part of its feasibility study for planned potable reuse, Tampa, Florida, conducted a "vulnerability assessment" to determine the suscep-

TABLE 2-4 Reductions in Los Angeles County
Wastewater Influent Loadings from 1975 to 1995

Constituent	Percent Reduction
Arsenic	68
Cadmium	92
Chromium	95
Copper	58
Lead	92
Mercury	46
Nickel	77
Zinc	73
Cyanide	95
Total identifiable chlorinated hydrocarbons	99

SOURCE: County Sanitation Districts of Los Angeles County, 1995.

tibility of the Hookers Point AWT facility to an upset due to an industrial
user's release of chemicals to the sanitary sewer system (CH$_2$M Hill, 1993).
Of the service area's 45 industrial users, 31 already had emergency provisions to divert a catastrophic chemical release away from the sewers. A
detailed evaluation was made of the 14 remaining industrial users with
floor-drain connections to the sewer system to determine the potential
for accidental spills. Evaluations were made of emergency response plans
and the potential effects of a chemical spill on the AWT facility. It was
concluded that the AWT facility was well protected from plant upset,
pass-through, or interference due to an accidental spill of chemicals at an
industrial user's facility. This type of vulnerability assessment can be
quite useful for gathering the information necessary for emergency con-
tingency planning as well as to provide safety assurance for the technical
operation of potable reuse. (Chapter 6 contains a discussion of treatment
plant reliability.)

Table 2-5 summarizes the concentrations of certain priority organic
pollutants following secondary biological treatment in four different municipal districts as of 1987: Washington, D.C.; Orange County, California;
Phoenix, Arizona; and Palo Alto, California. The Palo Alto system receives wastewater from a typical residential/commercial community as
well as from a major university and from several electronics industries.
Its wastewater contains high concentrations of chlorinated solvents. The
Orange County wastewater comes from a variety of industries as well as
municipal and commercial activities; it contains relatively high concentrations of petroleum-related chemicals, including various aromatic hy-

drocarbons (benzenes and naphthalenes). Phoenix, Arizona, represents another large municipality with a variety of commercial activities.

The wastewater from Washington, D.C., comes largely from residential and government-related activities, and the above-noted organic contaminants are relatively low in concentration. It should be noted that the concentrations listed for the Washington, D.C., water are for a blend of biologically treated municipal wastewater and Potomac River water. The concentrations would be higher for many of the contaminants in the wastewater itself.

Table 2-5 also contains a comparison of data from Orange County for two different time periods to illustrate the effects of using different biological treatment processes and of segregating wastewaters to reduce the industrial contribution to reclaimed water. During the first period noted, all wastewaters from Orange County were treated by trickling-filter biological treatment, and this water was used as the influent to Water Factory 21, the AWT system. During the second period, the wastewaters were treated by the activated sludge process, and segregation reduced the industrial contribution in the water sent to Water Factory 21. These changes significantly reduced the chemical oxygen demand (COD) of the treated wastewater and the concentrations of both chlorinated and unchlorinated benzenes and naphthalene. However, the concentrations of trihalomethanes, which normally result from water disinfection, increased. As anticipated, better AWT effluent quality was achieved for contaminants when their influent concentration was lower—confirming that virtually any effort to improve the quality of incoming wastewater will improve the treated water's quality.

MANAGING DISINFECTION BY-PRODUCTS IN REUSE SYSTEMS

Disinfection ranks as the most important single process for inactivating microorganisms in water and wastewater treatment. However, in some cases, reactions of disinfectants with organic and inorganic constituents in the source water can create potentially harmful (in some cases, carcinogenic) disinfection by-products (DBPs) (Bull and Kopfler, 1991; ILSI, 1995). The most common disinfectants are chlorine-based oxidants, but ozone and ultraviolet light are also used. Other disinfectants, such as gamma radiation, bromine, iodine, and hydrogen peroxide, have been considered for disinfection of wastewater, but they are not generally used because of economical, technical, operational, or disinfection efficiency considerations.

A limited number of DBPs are regulated or being considered for regulation. Chief among these are the trihalomethanes, haloacetic acids

TABLE 2-5 Average Concentrations of Selected Contaminants in
Municipal Wastewater Following Secondary Biological Treatment
(concentrations in µg/liter unless otherwise indicated)

Contaminant	Washington, D.C.[a]	Orange County Water District 1st Period	Orange County Water District 2nd Period	Phoenix[b]	Palo Alto
Total organic carbon (TOC), mg/liter	4.5	30	16	9	11
Total organic halides	85		131	87	192
Trihalomethanes					
$CHCl_3$	1.5	1.6	3.5	3.5	13
$CHBrCl_2$	<0.3	0.1	0.5	0.3	0.2
$CHBr_2Cl$	<0.2	0.2	0.7	0.2	0.1
$CHBr_3$		0.1	0.5	0.1	0.0
Total	2.0	2.0	5.2	4.1	13.3
Other Chlorinated Organics					
1,1,1-Trichloroethane	<0.2	4.7	4.8	1.4	65
Trichloroethylene	<0.1	0.9	1.1	0.4	25
Tetrachloroethylene	<0.8	0.6	3.6	1.7	44
Chlorobenzene		2.5	0.1		0.3
o-Dichlorobenzene	<0.05	2.4	0.7	2.4	2.7
m-Dichlorobenzene	0.08	0.7	0.2	0.4	3.6
p-Dichlorobenzene	<0.11	2.1	1.9	1.8	5.4
1,2,4-Trichlorobenzene	<0.02	0.5	0.3	0.4	11.3
Nonchlorinated Organics					
Toluene	<0.12				
Ethylbenzene	<0.02	1.4	0.04	0.2	0.03
o-Xylene	<0.04			0.4	
m-Xylene	<0.05		0.01	0.8	0.2
p-Xylene	<0.05		0.05	0.2	0.06
Naphthalene	<0.04	0.6	0.1	0.2	3.3

[a]After mixing with a 1:1 blend of Potomac River water and Blue Plains-treated effluent.
[b]Samples taken from spreading basins after secondary treatment.

SOURCE: Modified from California Department of Water Resources, 1987.

(HAAs), bromate, and haloacetonitriles. THMs and HAAs are the most
thoroughly studied and probably the dominant chlorinated DBPs that
form under "normal" disinfection conditions used to treat drinking water and municipal wastewater. Such compounds typically account for
between 30 and 50 percent of the total halogen incorporated into organic

compounds as a result of chlorination, although they may account for as little as 3 percent or as much as 80 percent of the total. Most of the remaining organic chlorine is thought to be incorporated into larger, as yet unidentified organic compounds. Many of these disinfection by-products remain poorly characterized toxicologically. The total concentration of chlorine and other halogens incorporated into organic compounds is collectively referred to as the "total organic halogen" (TOX) concentration.

When treated wastewater is used to augment potable water supplies, the two main DBP-related issues are (1) whether the wastewater contributes more precursors for formation of DBPs than does the conventional water source and (2) whether the wastewater provides precursors that lead to formation of DBPs different from those formed in potable water systems. Seemingly innocuous organic compounds subject to microbial degradation and found at higher concentrations in wastewater than in the general environment could contribute to the formation of high concentrations of certain by-products. More research is needed to identify such chemicals. For example, amino acids are a precursor of the very mutagenic compound 3-chloro-4-dichloromethyl-5-hydroxy-2(H)-furanone (otherwise known as "MX"; Horth et al., 1990). The concentrations of amino acids and other MX precursors commonly found in surface water sources are so low (measured in nanograms per liter) that significant MX formation is considered very unlikely (ILSI, 1996). However, higher concentrations of MX precursors in wastewater may allow higher concentrations of MX to be formed.

Table 2-6 illustrates the variability in DBPs produced in reclaimed water. The reclaimed water studied came from five wastewater plants in southern California using secondary biological treatment, nitrification, filtration, and chlorine disinfection (using a concentration × time value of 450 mg/liter × min). Two samples were collected from each plant. The first sample was collected from a point upstream of the chlorine addition point, representing the effluent of the tertiary filters prior to addition of the chlorine. The second sample was collected from the effluent of the chlorine contact chamber. The samples were analyzed for the indicated chlorination by-products, including HAAs, aldehydes, THMs, chloral hydrate (CH), and TOX. To provide some perspective, the table includes other standard water quality measurements as well.

Substantial levels of DBPs were formed in all chlorine contact chambers. The levels of THMs in the chlorinated samples ranged from a low of 35 µg/liter to a high of 86 µg/liter, with an average of 67 µg/liter. The levels of HAAs in the chlorinated samples ranged from a low of 99 µg/liter to a high of 262 µg/liter, with an average of 191 µg/liter. TOX values ranged from 432 to 785 µg/liter. While the THM levels would

TABLE 2-6 Results of Water Quality Parameters and Disinfection
By-Products Sampling Program

		Plant 1	
Analyte	Unit	Inf.	Eff.
General Physical Parameters			
Cl_2 dose	mg/liter	16	—
Ammonia-N	mg/liter as N	0.01	0.05
pH	—	7.1	6.6
TOC	mg/liter	7.0	6.3
DOC	mg/liter	5.5	5.4
Alkalinity	mg/liter as $CaCO_3$	100	75
UV_{254} (filtered)	cm^{-1}	0.108	0.072
UV_{254} (unfiltered)	cm^{-1}	0.116	0.078
TSS	mg/liter	13	2.0
Turbidity	NTU[a]	5.0	1.0
Disinfection By-Products			
THMs	µg/liter	0.9	35
HAAs	µg/liter	1	115
HANs	µg/liter	ND	14
CP	µg/liter	ND	ND
HKs	µg/liter	ND	8.4
CH	µg/liter	ND	44
CNCl	µg/liter	ND	1.9
Aldehydes	µg/liter	1	21
TOX	µg/liter	74	495

NOTE: ND = indicates not detected at the following limits: individual HAAs 1 µg/liter;
CNCl 0.5 µg/liter; individual HANs 0.5 µg/liter; CP 0.5 µg/liter; individual HKs 0.5 µg/
liter; CH 0.5 µg/liter. Acronyms are as follows: CH = chloral hydrate; CNCl = cyanogen
chloride; CP = chloropicrin; DOC = dissolved organic carbon; HAAs = haloactetic acids;
HANs = haloacetonitriles; HKs = haloketones; TOC = total organic carbon; THMs =
trihalomethanes; TSS = total suspended solids.

generally meet the current standards for drinking water of 100 µg/liter,
the levels found may be considered high if these reclaimed waters were
to be used to augment potable supplies. Unless they were reduced by
dilution in the receiving waters or by degradation, many would violate
the DBP MCLs proposed for the EPA interim drinking water standards
(80 µg/liter THMs and 60 µg/liter HAAs).

In another study, a survey of several wastewater treatment plants as
well as literature values, Hull and Reckhow (1993) found that the TOX
generated by chlorination of municipal wastewater after secondary biological treatment ranged from 50 to more than 1500 µg/liter when the
wastewater was chlorinated in the laboratory at the chlorine dose nor-

| Plant 2 | | Plant 3 | | Plant 4 | | Plant 5 | |
Inf.	Eff.	Inf.	Eff.	Inf.	Eff.	Inf.	Eff.
24	—	13	—	12	—	15	—
0.2	<0.01	<0.01	<0.01	1.0	<0.01	0.2	<0.01
7.5	7.0	7.3	7.0	6.7	6.3	7.5	7.5
7.3	7.3	6.7	6.8	10.9	8.4	8.0	7.8
7.3	6.9	6.6	6.5	8.4	8.2	6.4	6.3
160	125	180	160	145	95	235	240
0.136	0.085	0.123	0.076	0.114	0.071	0.122	0.085
0.140	0.095	0.128	0.077	0.125	0.076	0.154	0.116
0.8	2.4	1.0	0.6	5.2	1.4	3.2	1.6
0.6	3.0	0.6	0.3	3.0	0.5	1.5	2.3
1.9	68	0.7	83	3.2	86	1	62
4	247	2	231	6	262	1	99
ND	32	ND	27	ND	33	ND	19
ND	1.5	ND	1.2	ND	3.7	ND	1
ND	10.5	ND	9.7	ND	17	ND	6.5
0.8	54	ND	52	0.8	76	ND	57
ND	ND	ND	1.6	ND	3.9	ND	3.7
9	114	3	87	12	113	10	47
97	605	52	548	110	785	54	432

[a]NTU = nephelometric turbidity units.

SOURCE: Montgomery Watson, 1997. Personal communication from Rhodes Trussell and Issam Najm, Montgomery Watson, Pasadena, Calif.

mally applied at the corresponding full-scale facility. The median increase in TOX upon chlorination was between 100 and 200 µg/liter. In all cases, monochloramine is likely to have been the dominant oxidant present once the chlorine and wastewater mixed. The variability in organic halogen formation among the plants appeared to be strongly associated with the nature of the incoming organics. The greatest TOX formation occurred at a treatment plant dominated by wastewater from a paper mill. The formation of TOX was also influenced by the chlorine-to-ammonia ratio.

Hull and Reckhow (1993) also found more DOX (dissolved organic halides) formed per unit of dissolved organic carbon in the sample in secondary effluent than in untreated wastewater, based on an experiment in which all samples were chlorinated under standardized conditions expected to produce a high monochloramine concentration. The results suggest that secondary biological treatment leaves a mix of organic compounds in the water that is, on average, more susceptible to chlorine substitution than are the organics in untreated water.

Tertiary treatment of municipal wastewater reduces the potential for DBP formation. Based on ongoing work in the Montebello Forebay area of Los Angeles County, Leenheer (1996) reported that the yield of halogenated organics in a tertiary treated (secondary biological treatment, filtration, and chlorine disinfection) wastewater effluent is approximately an order of magnitude less than in chlorinated surface water containing organic compounds derived from terrestrial sources. This finding is probably more applicable to potable reuse of wastewater than are Hull and Reckhow's findings, because wastewater intended for potable reuse is typically subject to advanced treatment to remove organics and ammonia prior to chlorination, so that the oxidant would be free chlorine (as in Leenheer's study) rather than chloramines (as in Hull and Rechhow's study). As in reclaimed water, TOX production in natural water varies considerably depending on the nature of the dissolved organic carbon. Whether any differences in TOX formation between disinfected reclaimed water and other "natural" water sources reflect differences in the original mix of organics or in the processes that altered them before chlorination is not yet clear.

MIXTURES OF UNIDENTIFIED ORGANIC COMPOUNDS IN RECLAIMED WATER

The vast majority of the compounds in municipal wastewater cannot be identified at the molecular level. Under these circumstances, certain lumped parameters have been widely used to provide partial information about the mixture.

The broadest measures of organic molecules in water are total organic carbon (TOC) and dissolved organic carbon (DOC). These analyses, which measure all carbon-containing molecules, are traditionally used as indicators of potential contamination that may be associated with organic matter. DOC concentrations in wastewater that has been subjected to effective secondary treatment typically range from 5 to 15 mg/liter. TOC concentrations are strongly influenced by the efficiency with which biological solids are captured in the settling and/or filtration steps that follow biological treatment. They may range from 5 to 25 mg/liter in secondary treated municipal wastewater. Advanced treatment processes incorporating carbon filters and/or reverse osmosis can reduce concentrations to less than 1 mg/liter TOC in the finished water.

Other parameters that characterize the unidentified organic compounds in wastewater are more discriminating than TOC. Many of the approaches used to characterize such compounds were first developed to characterize "natural" organic matter (NOM)—that is, the complex mixture of organic compounds found in any natural aquatic system. Analytical approaches typically begin by broadly distinguishing between hydrophobic and hydrophilic compounds, then further subdividing these into acidic, basic, and neutral compounds. These separations are based on the pH-dependent affinities of the organic molecules for certain types of commercially available adsorbing resins. The placement of a NOM molecule in a particular category describes only its dominant characteristic; any given organic molecule may contain multiple and disparate subunits that may not be characterized by these separations.

Other descriptors widely used to characterize organics in natural waters include the organics' molecular weight distribution, elemental composition (e.g., percent C, N, O, H, P), functional group distribution (e.g., aromatic versus aliphatic carbon), and chemical or biochemical classification (e.g., carbohydrates, amino acids). In general, the more sophisticated and specific an analytical technique becomes, the smaller is the fraction of the total organic content of a natural water or wastewater that it can characterize. For instance, the characterization of organic matter into (predominantly) acidic, basic, or neutral fractions can be applied to virtually 100 percent of the organic compounds in a sample, whereas attempts to identify compounds that fit into specific chemical classes characterize only about 15 to 20 percent.

Usually, less than 1 percent of the organic carbon in the sample can be identified as specific compounds. In one of the most intensive characterization studies completed to date, Ding et al. (1996) identified approximately 10 percent of the dissolved organic matter in a highly treated wastewater as specific compounds; somewhat more than half of the identifiable mass was ethylenediaminetetracetic acid (EDTA). Table 2-7 shows the results of this research.

Concentration, Isolation, and Identification Methods

Considerable effort has been expended over the past two decades to identify and characterize the complex mixture of chemicals in pristine natural waters, municipal wastewaters, and treated wastewaters. Most of this effort has been directed at organic compounds. As explained in Chapter 4, concentration techniques are also used in the preparation of samples

TABLE 2-7 Positive and Tentative Identifications and Estimated
Concentrations of Organic Residues from GAC and Chlorinated
GAC Effluents

			Estimated Concentration (μg/liter)	
Peak	Positive and Tentative Identification or Compound Class	MW[a]	GAC	Cl-GAC
1	CH_3-CO-CH(OC$_3$H$_7$)$_2$ (methyl glyoxal)[b]	174	2.8	9.6
2	Aldehyde compound	N[e]	0.7	2.0
5	N,N,N-trimethylbenzeneamine[c]	135	0.9	—
6	Aldehyde compound	N	0.8	0.6
10	CH_3-(C$_4$H$_2$O)-COOC$_3$C$_7$ (methyl furylic acid)[b]	168	0.7	2.9
11	C_2H_5OCH$_2$CH$_2$CH$_2$-CH(OC$_3$H$_7$)$_2$	218	7.7	18
12	Butanedioic acid, dipropyl ester[c]	202	5.1	13
15	C_3H_5-CH(OC$_3$H$_7$)$_2$	172	1.1	1.1
17	$(C_3H_7O)_2$CH-CH(OC$_3$H$_7$)$_2$ (glyoxal)[b]	262	4.4	11
19	Aldehyde compound	N	0.8	1.4
20	Nitrilodiacetic acid (NDA), dipropyl ester[c]	217	0.7	1.8
22	Hexanedioic acid, dipropyl ester[c]	230	0.6	2.9
23	Diethoxypropoxydicarboxylic acid, dipropyl ester	276	1.6	2.7
24	Diethoxypropoxydicarboxylic acid, dipropyl ester	276	0.9	1.4
25	C_3H_7OOC-CH$_2$O-C$_3$H$_6$O-CH$_2$-COOC$_3$H$_7$	276	3.4	5.9
28	Nitrilotriacetic acid (NTA), tripropyl ester[c]	317	0.9	2.0
29	C_3H_7OOC-CH$_2$O-(C$_3$H$_6$O)$_2$-CH$_2$-COOC$_3$H$_7$	334	1.5	2.7
30	Diethoxypropoxydicarboxylic acid, dipropyl ester	334	1.2	1.8
31	Diethoxypropoxydicarboxylic acid, dipropyl ester	334	1.0	1.5
32	C_3H_7OOC-(C$_3$H$_6$)-C$_6$H$_4$-OCH$_2$-COOC$_3$H$_7$	322	1.4	—
33	C_3H_7OOC-(C$_3$H$_6$)-C$_6$H$_4$-OCH$_2$-COOC$_3$H$_7$	322	8.3	—
34	Naphthalenedicarboxylic acid, dipropyl ester	300	7.8	16
35	C_3H_7OOC-CH$_2$-(C$_4$H$_8$)-C$_6$H$_4$-OCH$_2$-COOC$_3$H$_7$	350	1.0	—
36	C_3H_7OOC-(C$_3$H$_6$)-(C$_3$H$_6$)-C$_6$H$_4$-OCH$_2$-COOC$_3$H$_7$	364	6.7	—
37	C_3H_7OOC-(C$_3$H$_6$)-C$_6$H$_4$-OC$_2$H$_4$-OCH$_2$-COOC$_3$H$_7$	366	1.7	—
38	C_3H_7OOC-(C$_3$H$_6$)-(C$_3$H$_6$)-C$_6$H$_4$-OCH$_2$-COOC$_3$H$_7$	364	11	—
39	C_3H_7OOC-(C$_4$H$_8$)-(C$_4$H$_8$)-C$_6$H$_4$-OCH$_2$-COOC$_3$H$_7$	392	1.0	—
40	C_3H_7OOC-(C$_4$H$_8$)-(C$_4$H$_8$)-C$_6$H$_4$-OCH$_2$-COOC$_3$H$_7$	392	1.5	—
41	Ethylenediaminetetraacetic acid (EDTA), tetrapropyl ester[c]	460	110	140
42	C_3H_7OOC-(C$_3$H$_6$)-(C$_3$H$_6$)-C$_6$H$_4$-OC$_2$H$_4$-OCH$_2$-COOC$_3$H$_7$	408	0.9	—
43	C_3H_7OOC-(C$_3$H$_6$)-(C$_3$H$_6$)-C$_6$H$_4$-OC$_2$H$_4$-OCH$_2$-COOC$_3$H$_7$	408	2.1	—
44	C_3H_7OOC-(C$_3$H$_6$)-C$_6$H$_3$Br-OCH$_2$-COOC$_3$H$_7$[d]	400	—	1.3
45	C_3H_7OOC-(C$_3$H$_6$)-C$_6$H$_3$Br-OCH$_2$-COOC$_3$H$_7$[d]	400	—	1.3
46	C_3H_7OOC-(C$_3$H$_6$)-C$_6$H$_3$Br-OCH$_2$-COOC$_3$H$_7$[d]	422	—	2.0
47	C_3H_7OOC-(C$_3$H$_6$)-C$_6$H$_3$Br-OCH$_2$-COOC$_3$H$_7$[d]	442	—	0.8
48	C_3H_7OOC-(C$_3$H$_6$)-(C$_3$H$_6$)-C$_6$H$_3$Br-OC$_2$H$_4$-OCH$_4$-COCOC$_3$H$_7$[d]	486	—	2.0

TABLE 2-7 Continued

NOTE: GAC = granular activated carbon; MW = molecular weight.

[a]Molecular weight calculated for the propylated derivative, obtained from *i*-butane chemical ionization mass spectrum.
[b]Chemical name in parentheses indicates the parent compound.
[c]A compound was considered positively identified if the spectra for the compound agreed with reference spectra from the National Biological Survey and EPA-National Institutes of Health libraries or if both its retention time and spectra agreed with those of a standard, reference compound.
[d]Brominated compounds only found in chlorinated GAC effluent.
[e]N—not identified.

SOURCE: Reprinted, with permission, from Ding et al., 1996. © 1996 by Springer-Verlag, New York.

for *in vitro* and *in vivo* toxicity testing of reclaimed water. Substantial advances in these techniques can be attributed to improvements in three areas:

1. concentrating the analytes—that is, selectively removing water from the sample so that the analytes' concentrations increase;
2. isolating the analytes—that is, selectively removing the analyte(s) from solution so that their concentration increases and substances that might interfere with the analysis are eliminated; and
3. developing new and improved analytical tools.

Although the quality of the reagents and instrumentation for carrying out these tasks has improved in recent years, the basic approaches have not changed. Rather, the techniques have been better integrated, making the distinctions among them less critical. The most advanced chemical characterization methods involve an extensive combination of concentration, isolation, and analytical steps.

Techniques for concentrating nonvolatile organics include low-temperature evaporation, membrane processes (reverse osmosis and nanofiltration), and adsorption/desorption using macroreticular resins as the adsorbent followed by elution by a solvent or by changing the pH. Liquid or gas chromatography is often used to further isolate analytes, followed by mass spectrometry for identification of specific compounds.

Techniques for concentrating chemical contaminants by removing water from the sample pose a notable challenge—that of achieving the required concentration without the concentrated compounds interacting and creating a solution or solid substantially different from the starting material. For example, when a sample is evaporated to dryness and is then re-contacted with a volume of water equal to the original volume, a portion of the solid material often fails to redissolve, at least within a time frame of hours to days. Even if some moisture is left in the sample, some solids can precipitate and interfere with the recovery of trace organic chemicals or with the subsequent isolation steps. Organic compounds are often co-precipitated with inorganic salts when water is removed from the sample. Therefore, attempts are usually made to remove the salts at various points in the concentration process.

When membrane processes are used to concentrate the organics, the key parameters that can be specified are the pore size and composition of the membrane being used. Because many organic compounds of interest are larger than dissolved inorganic species, membranes with appropriate pore sizes can partially overcome the problem of inorganic-organic interactions because they retain the organics while allowing most of the inorganics to pass through. The trade-off, of course, is that the membrane also fails to collect the lower molecular weight organics. Sun et al. (1995) reported that reverse osmosis (RO) could recover from 74 to 94 percent of the dissolved organic carbon in natural water samples, with the average recovery approaching 90 percent. They pretreated the water by passing it through a cation exchange resin to replace polyvalent cations with either H^+ or Na^+, thereby reducing the likelihood of coagulation or precipitation of the organics in the RO concentrate. The membrane retained the small, uncharged organic acids less successfully than it did the corresponding ionized species. Sun et al. reported that some of the remaining organic matter could be recovered by rinsing the membrane with dilute NaOH; this suggests that the organics had attached to the membrane surface as an acidic precipitate or by an adsorption reaction. Silica also precipitated in the RO unit in some applications, leading to a severe reduction in the flux of water through the membrane, but this had little effect on the recovery efficiency for dissolved organics.

Another common technique for concentrating organics from these types of samples is adsorption/desorption onto macroreticular resins. Different fractions of the organic mixture in the sample are sorbed depending on the hydrophobicity of the resin and the elution technique. Then different subfractions are collected depending on the chemistry of the eluent. Most inorganic salts are not sorbed by the resins; therefore, desalting of the column eluents is sometimes necessary. The technique was first popularized by a group of researchers at the U.S. Geological Survey in the late 1970s and early 1980s and has been used extensively since then (Aiken et al., 1979; Leenheer, 1981, 1984; Leenheer and Noyes, 1984; Thurman and Malcolm, 1981). Over the years, numerous modifications to the process have been suggested to improve the recovery efficiency of organic compounds and/or the ability to isolate various molecular groups. Aiken and Leenheer (1993) recently reviewed the technique and its capabilities, and Town and Powell (1993) examined some of its limitations.

The hydrophilic fraction of natural organics is particularly difficult to concentrate, isolate, and characterize because the inorganic salts in the sample are difficult to separate from the target species. Leenheer (1996) recently described an approach to overcome many of these problems by a combination of adsorption, elution, precipitation, and selective dissolution steps that separate the hydrophilic organics into neutral, acidic, and "ultra-acidic" fractions.

Once the organic compounds in a complex mixture have been separated and concentrated by one or a combination of the techniques noted above, the various fractions can be characterized by a number of analytical methods that focus on composite properties of the compounds. Composite properties that are frequently reported include elemental composition; molecular weight distribution; UV, IR, and ^{1}H- and ^{13}C-NMR (nuclear magnetic resonance) spectra of the samples; and GC-MS (gas chromatography-mass spectrometry) spectra of the fractions, sometimes after further processing, such as by pyrolysis.

Comparison of Wastewater to Natural Water

The physical, chemical, and biological processes that generate and modify organic compounds in wastewater and in natural systems share many similarities. For instance, the universality of basic metabolic pathways for the degradation of organic material ensures that most of the biologically generated organic matter from these different sources will have a great deal in common. Indeed, biological treatment processes in wastewater plants have been developed using natural systems as models, with the engineering aimed largely at compressing the time and volume required for the natural processes to occur. Many of the abiotic processes that remove organic molecules from solution in natural systems (chemical oxidation, photolysis, volatilization, and sorption) also have analogues in wastewater treatment systems. As a result, the chemical characteristics of wastewater-derived and naturally derived organic compounds probably overlap extensively.

Furthermore, after discharge to the ambient receiving water, organics of wastewater origin are gradually transformed into organics that more closely resemble natural compounds. Gray and Bornick (1996) used pyrolysis-GC-MS to analyze changes in the wastewater-derived mixture after passage through an artificial wetlands. They reported that over time, the organics in the mixture increasingly shifted toward a distribution characteristic of natural systems.

On the other hand, researchers have identified some important differences between natural and wastewater-derived organics. Some organic compounds found not at all or only at low concentrations in natural systems appear at much higher concentrations in wastewater even after it has been subjected to extensive treatment. Barber et al. (1996) illustrated the use of several of these organic compounds to determine the origins of organic contamination from municipal and industrial wastewater in the Mississippi River (Table 2-8).

In an attempt to identify distinctive features that could be used as indicators of wastewater contributions to a water sample, Peschel and Wildt (1988) compared various characteristics of treated wastewaters with those of natural waters. They reported that differences between wastewater organics and natural organic matter from the Ruhr River were too small to be useful for this purpose. On the other hand, Fujita et al. (1996) concluded that aggregate parameters were useful for following "longer-term processes involved in the turnover of organic carbon in aquifers" and that specific organic compounds (such as EDTA and alkylphenoxy ethoxycarboxylates, or APDCs) were useful as markers of wastewater. In a similar study, researchers analyzed ground water in the Montebello Forebay area, which is partially recharged with reclaimed water in spreading basins (Nellor et al., 1984). They found that aliphatic compounds comprised a substantially greater portion of the hydrophobic acids (the "fulvic acid" fraction) and that aromatic compounds existed in smaller proportions in wells containing reclaimed water than in wells less affected by human activity. The distinction apparently persists for several years after reclaimed water is introduced into the aquifer. This relationship probably derives from the fact that, in unperturbed environments, the precursor for much of the natural organic matter in water is thought to be lignin, a highly aromatic polymer with limited solubility that serves as a major structural component of plants. Further, natural organic matter in reclaimed water has been subjected to more microbial activity than natural organic matter in unperturbed water, and microbial activity increases its solubility. Presumably for the same reason, the specific ultraviolet absorbance (or absorbance per unit mass of organic carbon) is lower in wastewater than in natural water without substantial wastewater input (Debroux et al., 1996).

Risks of Nonionic Detergents in Reuse Systems

A group of nonionic detergents known as alkylphenylpolyethoxylates (or APnEOs, where n represents the number of ethoxy groups in

TABLE 2-8 Organic Compounds Measured to Evaluate Wastewater Contamination of the Mississippi River, 1987-1992

Contaminant	Abbreviation	Compounds and Sources
Dissolved organic carbon	DOC	All natural and synthetic organic compounds, regional-scale natural sources
Fecal coliform bacteria	None	Bacteria derived from human and animal fecal wastes; from sewage effluents and feedlot and agricultural runoff
Methylene-blue-active substances	MBAS	Composite measure of synthetic and natural anionic surfactants; predominantly from municipal sewage-wastewater discharges
Linear alkylbenzenesulfonate	LAS	Complex mixture of specific anionic surfactant compounds used in soap and detergent products; primary source is domestic sewage effluent
Nonionic surfactants	NP, PEG	Complex mixture of compounds derived from nonionic surfactants that includes nonylphenol (NP) and polyethylene glycol (PEG) residues; from sewage and industrial sources
Adsorbable organic halogen	AOX	Adsorbable halogen-containing organic compounds, including by-products from chlorination of DOC and synthetic organic compounds, solvents, and pesticides; from multiple natural and anthropogenic sources
Fecal sterols	None	Natural biochemical compounds found predominantly in human and livestock wastes; primary source is domestic sewage and feedlot runoff
Polynuclear aromatic hydrocarbons	PNA	Complex mixture of compounds, many of which are priority pollutants; from multiple sources associated with combustion of fuels

Table continues on next page

TABLE 2-8 Continued

Contaminant	Abbreviation	Compounds and Sources
Caffeine	None	Specific component of beverages, food products, and medications specifically for human consumption; most significant source is domestic sewage effluent
Ethylenediaminetetraacetic acid	EDTA	Widely used synthetic chemical for complexing metals; from a variety of domestic, industrial, and agricultural sources
Volatile organic compounds	VOCs	A variety of chlorinated solvents and aromatic hydrocarbons; predominantly from industrial and fuel sources
Semivolatile organic compounds	TTT, THAP	Wide variety of synthetic organic chemicals including priority pollutants and compounds such as trimethyltriazinetrione (TTT) and trihaloalkyl-phosphates (THAP); predominantly from industrial sources

SOURCE: Modified from U.S. Geological Survey, 1996.

the polymer) have received a great deal of attention lately because their breakdown products have been identified as potential hormone disruptors (U.S. EPA, 1997). Extremely low concentrations of these compounds have been shown to cause hormonal changes in fish; effects on humans are not yet established. Most types of household and industrial detergents contain a mixture of such compounds, with n values ranging from 1 to at least 18 and perhaps higher (Ahel et al., 1994b). Although the detergent molecules themselves are thought to be relatively innocuous, waste treatment breaks them down to smaller AP(nEO) compounds (almost all with n equal to 1 or 2), alkyl phenols (APs), and alkylphenylpoly-ethoxycarboxylates (AP(nEC)s), which are more toxic than the parent compounds (Ahel et al., 1994b).

Over the last two decades, Giger and various colleagues have extensively studied the fate of the detergent compounds and their metabolites in wastewater treatment systems and downstream (e.g., Ahel et al., 1994a, 1994b, 1996; Field et al., 1995; Giger et al., 1981, 1984). Detergent breakdown products resist further degradation and can accumulate in the environment. AP compounds, for instance, are hydrophobic and tend to accumulate in the sludge generated during waste treatment or to adsorb to organic matter in the receiving system (or in the soil if the wastewater is used for aquifer recharge via soil infiltration). AP(nEO) compounds are somewhat hydrophobic and may be present in either sludge or aqueous effluent, from which they are likely to be removed subsequently by sorption and/or biological degradation. AP(nEC) compounds, on the other hand, are hydrophilic and highly resistant to degradation, so they persist in the treated effluent far downstream of the discharge point. Fujita et al. (1996) found that AP(nEC) compounds persist through most tertiary waste treatment processes (lime addition and coagulation, rapid sand filtration, activated carbon adsorption, chlorination). That study also found, however, that these compounds might be altered by carboxylation of the alkyl group and, possibly, bromination of the aromatic ring. In addition, Fujitsu et al. (1996) found that reverse osmosis efficiently removed AP(nEC) compounds from the reclaimed water.

The AP(nEO)-AP(nEC) system (including the carboxylated and brominated derivatives) exemplifies the subtlety and complexity of the chemical/toxicological issues associated with using wastewater to augment potable water supplies. Wastewater often contains a potentially large number of chemicals and other environmental agents suspected of affecting human and animal endocrine systems in addition to breakdown products of nonionic detergents. General removal of TOC in advanced wastewater treatment systems would probably reduce concentrations of these compounds, but this issue has not been examined. As more organic chemicals are identified in wastewater at lower concentrations and as their biochemical effects are better understood, these health issues will arise more frequently. The development of rational approaches for understanding and reducing the associated risk from mixtures of organic chemicals should be a key goal for future research.

Use of Surrogate Parameters

The absence of a reliable technique for detecting a compound or quantifying its potential concentration in reclaimed water creates significant uncertainty regarding health risks to the water's consumers. The impossibility of identifying the complete mix of compounds present in a wastewater or a water supply source means that such uncertainty will remain a perpetual issue in evaluations of potable reuse of wastewater. This uncertainty could be partly reduced by a reliable method for toxicological testing that could establish a measure of safety even when individual contaminants cannot be identified. Chapters 4 and 5 discuss such toxicological testing issues.

Another approach is to establish a quantifiable limit of a surrogate or composite parameter that would provide some information on the concentration or behavior of unknown or suspected target compounds. The total organic carbon concentration, for instance, is widely used as a practical evaluator of water and wastewater treatment processes. Some would argue that a parameter as indiscriminating as TOC provides negligible value for indicating potential hazards associated with consumption of a water; this assertion is probably justified if one wishes to compare the risk associated with organics in different water sources, each of which contains several milligrams of TOC per liter. Further, from a strict public health perspective, removal of TOC is of limited objective value.

However, removing TOC from a water supply by any treatment process almost certainly reduces (though not necessary proportionally) the concentration of potentially hazardous, unidentified organic compounds. Diluting wastewater, as by discharging it into a receiving water, has a similar effect. Either method of reducing the contribution of treated wastewater to the DOC or TOC of a water source in a reuse situation might reduce user exposure to hazardous, unidentified wastewater compounds.

Other surrogate parameters may provide qualitative rather than quantitative information about unidentified organics. For instance, if two treated wastewaters contain equal concentrations of TOC, but one has a larger hydrophobic component and correspondingly greater value for specific ultraviolet light absorption, those differences undoubtedly reflect real and possibly important differences in the suites of organic compounds contributing to the TOC. Such analyses may make it possible to design an AWT process that targets classes of compounds specific to or particularly enriched in wastewater. Such a treatment process would convert the wastewater's population of organic molecules to one resembling naturally occurring populations. Unfortunately, such an approach begs the question of whether the distinctions being detected and reduced have significant health risks. For example, if the wastewater organics are relatively enriched in polysaccharides (a component of most foods), does it make sense to focus on polysaccharide removal simply to cause the mix of waste-derived organics to appear (and perhaps to be) more like natural organics?

Analytical chemistry techniques alone cannot address the question of risks from unknown organic chemicals. Toxicological methods must be used. However, the chemistry and toxicology can inform one another to identify the most promising and least promising areas of investigation. For example, the threat posed by endocrine disrupters was recently discovered when toxicological findings regarding abnormalities in fish inspired a search for possible explanations in wastewater effluents and receiving waters. That search identified several compounds that might be responsible, including both synthetic chemicals produced industrially and natural compounds produced as human metabolites. In the future, it seems increasingly likely that new compounds of concern will be identified by this sequence of events. Chapters 4 and 5 discuss in more detail such toxicological methods and issues relevant to potable reuse.

CONCLUSIONS

Municipal wastewater contains many chemicals that present known or potential health risks if ingested and that must be removed or reduced before such water is reused to augment a drinking water supply. Such chemical contaminants fall into three groups: (1) inorganic chemicals and minerals that are present naturally in the potable water supply; (2) chemicals created by industrial, commercial, and other human activities in the wastewater service area; and (3) chemicals that are added or generated during water and wastewater treatment and distribution processes. Any project to reclaim and reuse such water to augment drinking supplies must adequately account for all three categories of contaminants.

RECOMMENDATIONS

The recommendations below suggest several important guidelines for protecting against risks from chemical contaminants in potable reuse systems.

• **The research community should study in more detail the organic chemical composition of wastewater and how it is affected by treatment, dilution, soil interaction, and injection into aquifers.** Wastewater contains a much greater number of compounds than is covered by drinking water standards. The composition of wastewater and fate of the compounds it contains need to be better understood to increase the certainty that health risks of reclaimed water have been adequately controlled through treatment and transport and storage in the environment.

• **Projects proposing to use wastewaters as drinking water sources should document all major chemical inputs into the wastewater.** To the extent that domestic inputs to wastewaters can be assumed to be consistent throughout the United States, projects can estimate household chemical inputs on a per capita basis. For industrial inputs, projects should undertake a major effort to quantify the inputs of industrial chemicals, paying special attention to chemicals of greatest health concern.

• **For contaminants addressed by existing federal drinking water standards, reuse projects should bring concentrations within those standards' guidelines through a combination of source control within the service area, removal by secondary or tertiary waste treatment processes, dilution or removal in the receiving water, and removal in the drinking water treatment plant.** The regulations provided by the Safe Drinking Water Act and other federal guidelines cannot alone ensure the safety of drinking water produced from wastewater. However, for the contaminants they do address, those regulations are the best means available for judging the water's suitability for potable use.

• **The research community should determine whether chlorination of wastewater leads to formation of unique disinfection by-products or provides conditions that would lead to formation of higher levels of nonregulated but highly toxic by-products.** Whether reclaimed water forms significantly different by-products than natural waters upon disinfection is not yet clear. Clear guidance presently being developed for these common by-products in drinking water will make it easier to assess any threat they pose.

• **The risks posed by unknown or unidentifiable chemicals in reuse systems should be managed by a combination of reducing concentrations of general contaminant classes, such as total organic carbon, and conducting toxicological studies of the water.** Because it will never be possible to identify all the potentially harmful chemicals in treated wastewater, it will never be possible to definitively say the risk they pose has been reduced to acceptable levels. Nevertheless, in the absence of contravening data, one can generally assume that reducing the concentration of general categories of contaminants, such as TOC, also reduces risks posed by specific contaminants. If the proper controls and monitoring of wastewater inputs are in place, the health concerns associated with total organic carbon of wastewater origin should diminish as its overall contribution to the water supply diminishes. Implementation of these precautionary measures will reduce but not necessarily eliminate the need for toxicological studies and monitoring. Although establishing a TOC limit for potable reuse appears to be a legitimate risk management strategy, the nature of the organic carbon will influence what the limit should be. The committee believes this judgment should be made by local regulators, integrating all the information they have available to them concerning a specific project.

• **Finally, any reuse project should include a focused program for monitoring and ensuring the safety of the product water.** This program should be updated periodically as inputs to the system change or as its results reveal areas of weakness. Pretreatment requirements, wastewater treatment processes, and/or monitoring requirements may need to be modified to protect public health from exposure to specific chemical contaminants.

REFERENCES

Ahel, M., W. Giger, and M. Koch. 1994a. Behavior of alkylphenol polyethoxylate surfactants in the aquatic environment. I. Occurrence and transformation of in sewage treatment. Water Resources 28(5):1131-1142.

Ahel, M., W. Giger, and C. Schaffner. 1994b. Behavior of alkyphenol polyethoxlate surfactants in the aquatic environment. II. Occurrence and transformation in rivers. Water Research 28(5):1143-1152.

Ahel, M., C. Schaffner, and W. Giger. 1996. Behavior of alkyphenol polyethoxylate surfactants in the aquatic environment. II. Occurrence and elimination of their persistent metabolites during infiltration of river water to groundwater. Water Research 30(1):37-46.

Aiken, G. R., and J. A. Leenheer. 1993. Isolation and chemical characterization of dissolved and colloidal organic matter. Chemistry and Ecology 8:135-151.

Aiken, G. R., E. M. Thurman, R. L. Malcolm, and H. F. Walton. 1979. Comparison of XAD macroporous resins for the concentration of fulvic acid from aqueous solution. Analytical Chemistry 51(11):1799.

Barber, L. B., J. A. Leenheer, W. E. Pereira, T. I. Noyes, G. K. Brown, C. F. Tubor, and J. H. Writer. 1996. Organic contamination of the Mississippi River from municipal and industrial wastewater. P. 140 in Contaminants in the Mississippi River 1987-1999. Mead, R. H. (ed.) U.S. Geological Survey Circular 1133. Denver, Colo.: U.S. Geological Survey.

Bull, R. J., and F. C. Kopfler. 1991. Health Effects of Disinfectants and Disinfection By-Products. Denver, Colo.: American Water Works Association and AWWA Research Foundation.

California Department of Water Resources. 1987. Report of the Scientific Advisory Panel on Groundwater Recharge With Reclaimed Wastewater. Prepared for State of California, State Water Resources Control Board, Department of Water Resources, and Department of Health Services, November 1987. Sacramento, Calif.

CH$_2$M Hill. 1993. Tampa Water Resources Recovery Project Pilot Studies, Volume 1 Final Report. Tampa, Fla.: CH$_2$M Hill.

County Sanitation Districts of Los Angeles County. 1995. EPA National Pretreatment Program Excellence Awards Application, Whittier, Calif.

Debroux, J. F., G. Aiken, and G. Amy. 1996. The structure and character of organic matter in secondary effluents intended for potable reuse. Presented at the Workshop on the Influence of Natural Organic Matter Characteristics on Drinking Water Treatment and Quality, Poitiers, France, September 18 to 19.

Ding, W., Y. Fujita, R. Aeschimann, and M. Reinhard. 1996. Identification of organic residues in tertiary effluents by GC/EI-MS, GS/CI-MS, and GC/TSQ-MS. Fresenius Journal of Analytical Chemistry 354:48-55.

Field, J. A., T. M. Field, T. Poigner, H. Siegrist, and W. Giger. 1995. Fate of secondary alkane sulfonate surfactants during municipal wastewater treatment. Water Research 29(5):1301-1307.

Fujita, Y., W. Ding, and M. Reinhard. 1996. Identification of wastewater dissolved organic carbon characteristics in reclaimed wastewater and recharged groundwater. Water Env. Res. 68 (5):867-876.

Giger, W., E. Stephanou, and C. Schaffner. 1981. Persistent organic chemicals in sewage effluents: I. Identifications of nonylphenols and nonylphenolethoxylates by glass capillary gas chromatography/mass spectrometry. Chemosphere 10(11/12):1253-1263.

Giger, W., P. H. Brunner, and C. Schaffner. 1984. 4-Nonylphenol in sewage sludge: Accumulation of toxic metabolites from nonionic surfactants. Science 225:623-625.

Gray, K. A., and R. M. Bornick. 1996. Use of PY-GS-MS to characterize natural organic material in an artificial wetlands: Issues related to drinking water quality. Paper presented at the Workshop on the Influence of Natural Organic Matter Characteristics on Drinking Water Treatment and Quality, Poitiers, France, September 18 to 19.

Hannah, S. A. , B. M. Austern, A. E. Eralp, and R. H. Wise. 1986. Comparative removal of toxic pollutants by six wastewater treatment process. Journal of the Water Pollution Control Federation 58(1):27-34.

Horth, H., M. Fielding, H. A. James, M. J. Thomas, T. Gibson, and P. Wilcox. 1990. The production of organic chemicals and mutagens during chlorination of amino acids in water. Pp. 111-124 in Jolley, R. L. et al., (eds.) Water Chlorination: Chemistry, Environmental Impact, and Health Effects. Vol. 6. Lewis, Chelsea, MI.

Hull, C. S., and D. A. Reckhow. 1993. Removal of DOX and DOX precursors in municipal wastewater treatment plants. Water Research 27(3):419-425

ILSI. 1995. Disinfection by-products in drinking water: Critical issues in health effects research. Pp. 110-120 in Workshop Report. Chapel Hill, N.C. October 23 to 25.

ILSI Risk Science Institute Pathogen Risk Assessment Working Group. 1996. A conceptual framework to assess the risks of human disease following exposure to pathogens. Risk Analysis 16(6): 841-848.

Lauer, W. C. 1990. Denver's Direct Potable Water Reuse Demonstration Project, Final Report to EPA. Denver, Colo.: Denver Water Department.

Lauer, W. C., S. E. Rogers, A. M. La Chance, and M. K. Nealy. 1991. Process selection for potable reuse health effects studies. Journal of the American Water Works Association 83:52-63.

Leenheer, J. A. 1981. Comprehensive approach to preparative isolation and fractionation of dissolved organic carbon from natural waters and wastewaters. Environmental Science Technology 15(5):578.

Leenheer, J. A. 1984. Concentration, partitioning, and isolation techniques. Pp. 84-166 in Minear, R. A., and L. H. Keith (eds.) Water Analysis. Vol. III. Orlando, Fla.: Academic Press, Inc.

Leenheer, J. A., and T. I. Noyes. 1984. A filtration and column-adsorption system for onsite concentration and fractionation of organic substances from large volumes of water. U.S. Geological Survey Water Supply Paper 2230.

Mills, W. R. 1996. 1995 Annual Report, Orange County Water District Wastewater Reclamation, Talbert Barrier, and Recharge Project. Prepared for the California Regional Water Quality Control Board, Santa Ana Region. Report Order No. 91-121.

Nellor, M. H., R. B. Baird, and J. R. Smyth. 1984. Health Effects Study Final Report, March 1984. Whittier, Calif.: County Sanitation Districts of Los Angeles County.

Nellor, M. H., R. B. Baird, and J. R. Smyth. 1985. Health Effects of Indirect Potable Water Reuse. Journal of the American Water Works Association July:88-96.

Peschel, G., and T. Wildt. 1988. Humic substances of natural and anthropogenic origin. Water Research 22:105-108.

Petrasek, Jr., A. C., I. J Kugelman, B. M . Austern, T. A. Pressley, L. A. Winslow, and R. H. Wise. 1982. Fate of toxic organic compounds in wastewater treatment plants. Journal of the Water Pollution Control Federation 55:1286-1296.

Richards, D. J., and W. K. Shieh. 1986. Biological fate of organic priority pollutants in the aquatic environment. Water Research 20:1077-1090,

Sun, L., E. M. Perdue, and J. F. McCarthy. 1995. Using reverse osmosis to obtain organic matter from surface and ground waters. Water Research 29(6): 1471-1477.

Thurman, E. M., and R.L. Malcolm. 1981. Preparative isolation of humic substances. Environmental Science and Technology 15(4):463.

Town, R. M., and H. K. J. Powell. 1993. Limitations of XAD resins for the isolation of the non-collodial humic fraction in soil extracts and aquatic samples. Analytica Chimica Acta 271:195-202.

U.S. Environmental Protection Agency. 1997. Special Report on Environmental Endocrine Disruption: An Effects Assessment and Analysis. Report No. EPA/630/R-96/012, February. Washington, D.C..: EPA.

U.S. Geological Survey. 1996. Meade, R. H. (ed.) Contaminants in the Mississippi River, 1987-92. USGS Circular 1133. USGS, Denver, Colorado.

Water Replenishment District of Southern California. 1996. Annual Report on Results of Water Quality Monitoring: Water Year 1995-1996. Ceritos, Calif.: Water Replenishment District of Southern California.

Western Consortium for Public Health. 1997. Total Resource Recovery Project Aqua II San Pasqual Health Effects Study, Final Summary Report. Prepared for the City of San Diego Water Utilities Department. Oakland, Calif.: Western Consortium for Public Health.

3

Microbial Contaminants in Reuse Systems

Traditionally, bacterial and other indicators have been used to evaluate the effectiveness of water and wastewater treatment systems in inactivating microorganisms. Except for special studies, relatively little occurrence information is available for the pathogens that actually pose health risks. Over the past few years, however, renewed attention has been given the health risks from microbial contamination of drinking water, and nationwide monitoring programs are being instituted. In the meantime, much of the information available on specific pathogens comes from microbial monitoring and studies of nonpotable and some potable reuse projects. Knowing the occurrence and concentration of specific pathogens in reclaimed water is critical to determining exposure and thus assessing the potential health risks of potable water reuse.

WATERBORNE DISEASES

Microorganisms associated with waterborne disease are primarily enteric pathogens, which have a fecal-oral route of infection (either human-to-human or animal-to-human) and survive in water. These bacteria, viruses, and protozoa can be transmitted by consumption of fecal-contaminated water, but they can also be spread through person-to-person contact, contaminated surfaces, and food. Any potable water supply receiving human or animal wastes can be contaminated with microbial agents. Even pristine water supplies have been associated with disease

TABLE 3-1 Common Infectious Agents Potentially Present in Untreated Municipal Wastewater

Agent	Disease
Protozoa	
Entamoeba histolytica	Amebiasis (amebic dysentery)
Giardia lamblia	Giardiasis
Balantidium coli	Balantidiasis (dysentery)
Cryptosporidium	Cryptosporidiosis, diarrhea, fever
Helminths	
Ascaris (roundworm)	Ascariasis
Trichuris (whipworm)	Trichuriasis
Taenia (tapeworm)	Taeniasis
Bacteria	
Shigella (4 spp.)	Shigellosis (dysentery)
Salmonella typhi	Typhoid fever
Salmonella (1700 serotypes)	Salmonellosis
Vibrio cholerae	Cholera
Escherichia coli (enteropathogenic)	Gastroenteritis
E. coli 0157:H7 (enterohemorrhagic)	Bloody diarrhea
Yersinia enterocolitica	Yersiniosis
Leptospira (spp.)	Leptospirosis
Legionella pneumophila	Legionnaire's disease, Pontiac fever
Campylobacter jejuni	Gastroenteritis
Viruses	
Enteroviruses (72 types)	
Poliovirus	Paralysis, aseptic meningitis
Echovirus	Fever, rash, respiratory illness, aseptic meningitus, gastroenteritis, heart disease
Coxsackie A	Herpangina, aseptic meningitus, respiratory illness
Coxsackie B	Fever; paralysis; respiratory, heart, and kidney disease
Norwalk	Gastroenteritis
Hepatitis A virus	Infectious hepatitis
Adenovirus (47 types)	Respiratory disease, eye infections
Rotavirus (4 types)	Gastroenteritis
Parvovirus (3 types)	Gastroenteritis
Reovirus (3 types)	Not clearly established
Astrovirus (7 types)	Gastroenteritis
Calicivirus (2-3 types)	Gastroenteritis
Coronavirus	Gastroenteritis

SOURCE: Adapted from Hurst et al., 1989; Sagik et al., 1978.

outbreaks, presumably due to *Giardia* contamination from wildlife in the watershed.

Table 3-1 shows the bacteria, viruses, and protozoan parasites potentially present in untreated municipal wastewater. Wastewater may also contain helminths (intestinal worms), but these waterborne parasites will not be discussed in this report because conventional wastewater treatment removes helminths and their relatively large ova and cysts. Other microorganisms, such as *Legionella*, are sometimes classified as waterborne disease agents but will not be addressed because their airborne routes of transmission are distinctly different from the transmission routes of enteric microbial agents.

Concerns over particular waterborne microorganisms have changed over the years due to improved sanitary conditions, the use of preventive medicine, and improved microbiological and epidemiological methods for identifying the microorganisms responsible for outbreaks. Microorganisms were first identified as agents of waterborne disease during the cholera outbreak in England in the 1860s. In the 1920s, typhoid fever was linked to the waterborne bacterium *Salmonella typhi*. *Giardia*, a water-

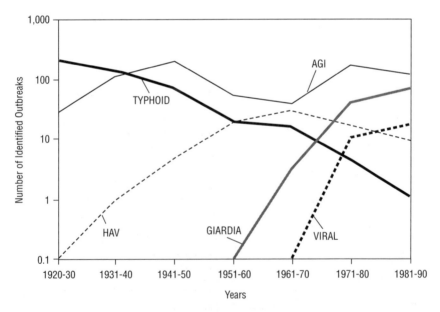

FIGURE 3-1 Changing trends in waterborne diseases in the United States in the twentieth century. NOTE: AGI = acute gastrointestinal illness of unknown etiology; HAV = hepatitis A virus.

borne protozoan, rose as a major concern in the 1960s; rotavirus and Norwalk virus were associated with a large number of outbreaks beginning in the 1970s; and *Cryptosporidium*, also a protozoan, was first associated with waterborne outbreaks in the 1980s (Figure 3-1).

Much of the information on the etiology of waterborne disease comes from investigations of outbreaks by state and local health departments and from voluntary reporting by physicians to the surveillance program maintained by the Centers for Disease Control and Prevention (CDC) and the Environmental Protection Agency (EPA). When an outbreak occurs and waterborne pathogens are suspected, epidemiological studies are conducted to identify whether water is the vehicle of transmission. If possible, the etiologic agent is determined by detection in clinical specimens collected from outbreak victims. For gastrointestinal illness, routine stool examinations by hospital laboratories typically include culturing for *Salmonella, Shigella,* and *Campylobacter* bacteria. At the specific request of a physician, many laboratories can also test for rotavirus, *Giardia*, and *Cryptosporidium*. Nevertheless, no specific agent is identified in many outbreaks, leaving the cause classified only as acute gastrointestinal illness of unknown etiology (AGI). Before 1982, in fact, most waterborne outbreaks reported were listed as AGI. Poor collection of clinical and/or water samples and limitations of diagnostic techniques for many enteric pathogens can prevent accurate determination of the pathogen. Clinical symptoms suggest that many of the AGI outbreaks may be due to viral agents, such as Norwalk virus and related human caliciviruses.

Diseases From Enteric Bacteria

Enteric bacteria are associated with human and animal feces and may be transmitted to humans through fecal-oral transmission routes. Most illnesses due to enteric bacteria cause acute diarrhea, and certain bacteria tend to produce particularly severe symptoms. As measured by hospitalization rates during waterborne disease outbreaks (that is, the percentage of illnesses requiring hospitalization), the most severe cases are due to *Shigella* (5.4 percent), *Salmonella* (4.1 percent), and pathogenic *Escherichia coli* (14 percent) (Gerba et al., 1994). There is now evidence suggesting that *Campylobacter, Shigella, Salmonella*, and *Yersinia* may also be associated with illness that causes arthritis in about 2.3 percent of cases (Smith et al., 1993).

Most *E. coli* are common, harmless bacteria found in the intestinal tracts of humans and animals, but some forms of *E. coli* are pathogenic and cause gastroenteritis. A particular strain, *E. coli* O157:H7, is enterohemorrhagic (causes bloody diarrhea), and 2 to 7 percent of infections have resulted in hemolytic uremic syndrome (HUS), in which red blood

TABLE 3-2 Waterborne Bacterial Agents of Concern

Bacteria	Average Reported Cases in the United States	Annual Case-Fatality Rate (%)[a]	Percent Waterborne[b]
Campylobacter	8,400,000	0.1	15
Pathogenic *Escherichia coli*	2,000,000	0.2	75
Salmonella nontyphoid	10,000,000	0.1	3
Shigella	666,667	0.2	10
Yersinia	5,025	0.05	35

[a]The number of deaths per case expressed as a percentage and based on total cases and deaths reported annually to the CDC.
[b]Percentage of cases attributed to water contact or water consumption.

SOURCE: Reprinted by permission of Elsevier Science from Bennett et al., 1987. ©1987 by American College of Preventive Medicine.

cells are destroyed and the kidneys fail. HUS has one of the highest mortality rates of all waterborne diseases. The microbial reservoir for *E. coli* O157:H7 appears to be healthy cattle, and transmission can occur by ingestion of undercooked beef or raw milk as well as by contaminated water. Two waterborne outbreaks of *E. coli* O157:H7 have been reported in the United States (CDC, 1993). Drinking water was associated with an outbreak of *E. coli* O157:H7 involving 243 cases, 32 hospitalizations, and 4 deaths in a Missouri community in 1989. Unchlorinated well water and breaks in the water distribution system were considered to be contributing factors. The other waterborne outbreak of *E. coli* O157:H7 involved 80 cases in Oregon in 1991 and was attributed to recreational water contact in a lake (Oregon Health Division, 1992). Prolonged survival of *E. coli* O157:H7 in water has been reported by Geldreich et al. (1992), who observed only a 2 log (99 percent) reduction after 5 weeks at 5°C.

Classical waterborne bacterial diseases such as dysentery, typhoid, and cholera, while still very important worldwide, have dramatically decreased in the United States since the 1920s (Craun, 1991). However, *Campylobacter*, nontyphoid *Salmonella*, and pathogenic *Escherichia coli* have been estimated to cause 3 million illnesses per year in the United States (Bennett et al., 1987). Hence, enteric bacterial pathogens remain an important cause of waterborne disease in the United States. Table 3-2 shows the number of enteric bacterial illnesses, the case-fatality rate reported annually from all cases, and the percentage of illnesses attributed to contaminated water supplies, which ranges from 3 to 75 percent. Enteric bacteria caused 14 percent of all waterborne disease outbreaks in the United States from 1970 to 1990 (Craun, 1991).

Diseases From Enteric Protozoa

The enteric protozoan parasites produce cysts or oocysts that aid in their survival in wastewater. Important pathogenic protozoa include *Giardia lamblia, Cryptosporidium parvum,* and *Entamoeba histolytica.* (Helminth ova are present in untreated wastewater; however, they are relatively large and tend to drop out of effluent after primary and secondary treatment.) Waterborne outbreaks of amebic dysentery, caused by *Entamoeba,* have not been reported in the United States in over 15 years (Bennett et al., 1987). *Giardia* is recognized as the most common protozoan infection in the United States and remains a major public health concern (Craun, 1986; Kappus et al., 1992). The reported incidence of waterborne giardiasis has increased in the United States since 1971 (Craun, 1986). An average of 60,000 cases are reported annually, and 60 percent are estimated to be waterborne (Bennett et al., 1987). Because *Giardia* is endemic in wild and domestic animals, infection can result from water supplies that have no wastewater contribution. Densities of *Giardia* cysts in untreated wastewater have been reported as high as 3375 per liter (Sykora et al., 1991).

TABLE 3-3 Illness Rates From Enteric Viruses

Virus Group	Annual Reported Cases in the United States[a]	Case-Fatality Rate (%)	Morbidity Rate (%)
Enteroviruses	6,000,000	0.001	Not known
Poliovirus	7	0.90	0.1-1
Coxsackievirus A	Not known	0.50	50
Coxsackievirus B	Not known	Not known	0.59-0.94
Echovirus	Not known	Not known	50
Hepatitis A virus	48,000	0.6	75
Adenovirus	10,000,000	0.01	Not known
Rotavirus	8,000,000	0.01	56-60
Norwalk agent	6,000,000	0.0001	40-59

[a]Cases reported to the CDC in 1985.

SOURCE: Bennett et al., 1987; Gerba and Rose, 1993.

TABLE 3-4 Emerging and Potential Waterborne Enteric Pathogens

Microorganism	Description	Clinical Syndrome
Calicivirus	Group of "small round structured viruses" approx. 27-35 nm diameter, SS[a] RNA. Includes Norwalk virus, Snow Mountain virus, and Hawaii virus	Acute gastroenteritis, major cause of outbreaks of nonbacterial, acute gastroenteritis
Astrovirus	Small, round structured virus approx. 28-30 nm diameter, SS RNA, 7 serotypes	Acute gastroenteritis, mainly in children and the elderly
Enteric adenovirus	Approx. 70-80 nm diameter, DS DNA[b] virus, mainly serotypes 40 and 41	Gastroenteritis with duration of 7-14 days; associated with 5-12% of pediatric diarrhea
Enteric coronavirus	Between 100 and 150 nm diameter, enveloped SS RNA virus; major gastrointestinal pathogens of animals, putative enteric pathogens for humans	Acute gastroenteritis
Torovirus	Enveloped. Approx. 100-150 nm diameter, SS RNA viruses; well-established enteric pathogens for animals, putative enteric pathogens for humans	Acute gastroenteritis
Picornavirus	Approx. 25-30 nm diameter, double-stranded RNA viruses	Diarrhea
Pestivirus	Single-stranded RNA viruses	Pediatric diarrhea
Helicobacter pylori	Typically, curved, gram-negative rods 3×0.5 μm, microaerophilic	Colonization of stomach causes persistent low-grade gastric inflammation; chronic infections may result in peptic ulcers and gastric cancer

Evidence of Waterborne Transmission	Reports of Occurrence	References
Numerous reports of waterborne outbreaks	Methods to detect in water are currently being developed	Kapikian et al., 1996
Waterborne outbreaks have been reported	No methods to detect in water	Matsui and Greenberg, 1996
None, but known to have fecal-oral transmission	Has been recovered from sewage	Foy, 1991 Petric, 1995
None, but epidemiologic evidence of fecal-oral transmission	No methods to detect in water	McIntosh, 1996
None	No methods to detect in water	Koopmans et al, 1991, 1993
None	No methods to detect in water	Pereira et al., 1988
None	No methods to detect in water	Yolken et al., 1989
Probable fecal-oral transmission; some epidemiologic studies have implicated type of water supply as an important risk factor	Lab studies demonstrated *H. pylori* survival for 10 days in freshwater; also evidence of prolonged survival as viable, nonculturable coccoid bodies. Recent report of PCR method to detect *H. pylori* in water[c]	Enroth and Engstrand, 1995 Shahamat et al., 1989 West et al., 1990

Table continues on next page

TABLE 3-4 Continued

Microorganism	Description	Clinical Syndrome
Cyclospora cayetanensis	Protozoa with oocysts 8-10 μm in diameter	Prolonged, self-limited diarrhea with average duration of 30 days

[a]SS RNA = single-strand RNA.
[b]DS DNA = double-strand DNA.
[c]PCR = polymerase chain reaction.

Giardia has also been detected in treated effluent and is much more resistant to disinfection with chlorine than bacteria.

Cryptosporidium was first described as a human pathogen in 1976. Cryptosporidiosis causes severe diarrhea; no pharmaceutical cure exists. Average infection rates in the United States, as measured by oocyst excretion in a population, have ranged from 0.6 to 20 percent (Fayer and Ungar, 1986). The disease can be particularly hazardous for people with compromised immune systems (Current and Garcia, 1991). Since 1985, seven reported waterborne outbreaks of cryptosporidiosis have occurred in the United States (Lisle and Rose, 1995). In 1993, *Cryptosporidium* was responsible for the largest waterborne disease outbreak ever recorded in the United States, causing approximately 400,000 illnesses in Milwaukee, Wisconsin. This outbreak was attributed to contamination of the surface water supply by either animal or human wastes (MacKenzie et al., 1995). All research to date suggests that the current standards for water chlorination are inadequate for inactivation of *Cryptosporidium* oocysts (Korick et al., 1990; Peeters et al., 1989). *Cryptosporidium* oocysts have been detected in municipal wastewater, but their concentrations and removal by wastewater treatment processes have not been fully evaluated (Madore et al., 1987; Rose et al., 1996; Villacorta-Martinez et al., 1992).

Diseases From Enteric Viruses

The enteric viruses are obligate human pathogens, which means they replicate only in the human host. Viruses are the smallest pathogenic agents. Their simple structure of a protein coat surrounding a core of

Evidence of Waterborne Transmission	Reports of Occurrence	References
Epidemiologic case-control study in Nepal implicated consumption of untreated water as a risk factor; 1990 outbreak in Chicago associated with rooftop water storage tanks	Methods to detect in water are currently under development	Ortega et al., 1993 Shlim et al., 1991

genetic material (DNA or RNA) allows prolonged survival in the environment. There are more than 120 identified human enteric viruses. Some of the better described viruses include the enteroviruses (polio-, echo-, and coxsackieviruses), hepatitis A virus, rotavirus, and Norwalk virus. Most enteric viruses cause gastroenteritis or respiratory infections, but some may produce a range of diseases in humans, including encephalitis, neonatal disease, myocarditis, aseptic meningitis, and jaundice (Gerba et al., 1995, 1996; Wagenkneckt et al., 1991; see Table 3-1). Cases of poliovirus are low in the United States due to almost universal immunization. Table 3-3 shows the average number of viral illnesses that occur annually in the United States for the different enteric viral groups. No general estimates exist regarding the percentage of viral illnesses attributable to contaminated water supplies.

Norwalk and Norwalk-like viruses cause most waterborne viral diseases. Norwalk virus usually causes mild diarrhea that lasts on average for two days. A significant portion of the waterborne outbreaks reported as AGI are probably caused by Norwalk-like viruses that are not identified because of diagnostic limitations; Kaplan et al. (1982) suggested that such viruses may cause 23 percent of all waterborne outbreaks reported as AGI. From 1989 to 1992, contaminated drinking water was implicated in four outbreaks associated specifically with Norwalk-like viruses and hepatitis A virus (Herwaldt et al., 1992; Moore et al., 1993). During the same period, 37 waterborne outbreaks of AGI affected 15,769 people. In 85 percent of the outbreaks, the water quality met national drinking water standards for coliform bacteria.

Emerging and Unknown Waterborne Pathogens

One concern about potable reuse of reclaimed water is the potential health risk from little-known or unknown pathogens. In more than half of all reported outbreaks of waterborne disease, no etiologic agent was ever determined. Some outbreaks that were thoroughly investigated suggest the existence of unrecognized pathogens. For example, "Brainerd diarrhea," first described in an outbreak in Brainerd, Minnesota, in 1983 (Osterholm et al., 1986), is characterized by chronic diarrhea lasting an average of 12 to 18 months. Similar symptoms were noted in several subsequent outbreaks in seven other states where the disease etiology was associated with poor-quality or untreated drinking water (Parsonnet et al., 1989). Intense microbiological analyses failed to identify any etiologic agent for this syndrome.

"Emerging" infectious diseases have been defined as those whose incidence in humans has increased within the past two decades or threatens to increase in the near future (Institute of Medicine, 1992). Some infectious agents, such as *Cryptosporidium*, were first described in the past 10 to 20 years but have more recently emerged as major causes of waterborne disease. Drinking water from potable reuse systems may pose a risk of exposure to emerging enteric pathogens because raw wastewater contains many enteric pathogens, the removal of which by treatment processes can only be inferred by other measures of microbial quality. The occurrence and health significance of many of these agents in finished drinking water are currently unknown.

Table 3-4 summarizes information on a number of recently recognized enteric pathogens known to have waterborne transmission or to have the potential for waterborne transmission via fecal-contaminated water. The table includes the sizes of these organisms (when known), since this may be relevant to their removal by specific water and wastewater treatment processes. (Several emerging enteric waterborne pathogens that are important outside the United States e.g., hepatitis E virus, group B rotavirus, and *Vibrio cholerae* O139 are not discussed here because these infections have not been transmitted within the United States.)

Norwalk virus and related human caliciviruses are considered emerging pathogens because new diagnostic techniques have recently identified their roles as major waterborne and foodborne pathogens. A number of other viruses are known or putative enteric pathogens. However, little or no evidence exists regarding waterborne transmission of these organisms. Methods to detect them in water and wastewater have not been developed, and little or no information exists about their survival or transmission in water. For instance, astroviruses are recently recognized enteric pathogens. Initially there were only a few anecdotal reports of transmission by contaminated water in the literature (Kurtz and Lee, 1987). More recently, large outbreaks and the importance of astroviruses have been recognized, and evidence for waterborne transmission is mounting (Abad et al., 1997). Enteric adenoviruses (serotypes 40 and 41, also known as subgenus F) are DNA viruses that are a common cause of pediatric diarrhea. Although adenoviruses have been recovered from sewage (Foy, 1991), there has been no evidence of drinking water waterborne transmission, though recreational outbreaks have been reported (Crabtree et al., 1997).

Coronaviruses were first observed in feces of persons with gastroenteritis by electron microscopy in 1975, but since then they have also been frequently detected in the feces of healthy people; their etiologic role in human diarrhea remains doubtful. Epidemiologic evidence suggests that fecal-oral transmission and personal hygiene may be key factors in transmission since several studies have noted that the highest prevalence rates were among populations with poor personal hygiene (Caul, 1994).

Toroviruses, which are well-established enteric pathogens of cattle and horses, have been found in stool samples from children and adults with diarrhea (Koopmans et al., 1991, 1993) but have remained unconfirmed as human pathogens. Similarly, picornavirus and pestivirus have been detected in fecal specimens from adults and children with diarrhea, but their clinical significance is not known.

The pathogenic bacterium *Helicobacter pylori,* formerly referred to as *Campylobacter pylori*, causes indigestion and abdominal pain, and chronic infection may result in peptic ulcers and gastric cancer. *H. pylori* infections occur throughout the world, and the prevalence of infection increases with age. Fecal-oral transmission of *H. pylori* infection has been suggested by several studies that implicated crowding, socioecomonic status, and consumption of raw, sewage-contaminated vegetables as risk factors for infection (Hopkins et al., 1993; Mendall et al., 1992; Mitchell et al., 1992). Studies in Peru have identified type of water supply (municipal vs. community wells) as a risk factor for infection with *H. pylori* and found that water source appeared to be a more important risk factor than socioeconomic status; children from high-income families who received their water from the Lima municipal water supply, which comes from a surface water source, were 12 times more likely to become ill than high-income children who drank well water, with community wells posing a higher risk than treated municipal supplies (Klein et al., 1991). Yet a seroprevalence survey of 245 healthy children in Arkansas found no relation between *H. pylori* seropositivity and type of water supply (municipal or well) (Fiedorek et al., 1991). However, the levels of fecal contamina-

Channels containing water hyacinths, used to provide secondary treatment at San Diego's Aqua II pilot facility. Photo by Joe Klein.

tion in the Peruvian water sources were likely substantially higher than those in Arkansas.

Cyclospora cayetanensis, formerly called "cyanobacterium-like bodies" (CLB) or "big *Cryptosporidium*," was identified as a new protozoan pathogen of humans in 1993 (Ortega et al., 1993). The organism causes persistent diarrhea. Outbreaks and case reports of *Cyclospora* diarrhea have been associated with exposure to fecal-contaminated water and fruit that may have been irrigated with fecal-contaminated water (Hale et al., 1994; Shlim et al., 1991). *Cyclospora* has also been implicated in waterborne transmission, but not as often as *Cryptosporidium* and *Giardia*. In June 1994, several cases of diarrhea were detected among British soldiers and dependents stationed in a small military detachment in Pokhara, Nepal (Rabold et al., 1994). The drinking water for the camp was a mixture of river and municipal water that was treated by chlorination. A candle filtration system was also used to remove particles but was not guaranteed to filter *Cyclospora*-sized particles. *Cyclospora* was detected in 75 percent of the diarrhea samples examined, and a water sample taken from the camp contained *Cyclospora* oocysts. Twenty-one cases of prolonged diarrhea in employees and staff physicians were noted on July 9, 1990, in a Chicago hospital (Huang et al., 1995). Upon investigation,

Cyclospora oocysts were identified in the stools. Epidemiological investigations implicated the tap water in the physicians' dormitory and indicated that the storage tank may have been contaminated. Although this outbreak has been identified as a waterborne outbreak, a plausible scenario for the contamination of the water has not been developed.

Another set of emerging protozoan pathogens is microsporidia—a general term that describes a large group of primitive, obligate, intracellular protozoa. Most reported cases of microsporidial infections have occurred among persons infected with human immunodeficiency virus (HIV) (Bryan et al., 1991), and recent epidemiological studies suggest that one species, *Enterocytozoon bieneusi*, is an important cause of chronic diarrhea in patients with AIDS (Weber et al., 1992). The development of methods to detect *H. pylori*, *Cyclospora cayetanensis* oocysts, and microsporidia spores in water and wastewater and to evaluate the risks associated with waterborne exposure to these pathogens was recently identified as a priority research need by the EPA (U.S. EPA, 1996).

A final emerging waterborne pathogen is *Isospora belli*, a protozoan that has been associated with one documented waterborne disease outbreak in troops in Panama (Goodgame, 1996). The oocyst is large (20 to 30 µm) and, while it is endemic in Africa, Asia, and South America, it is extremely rare in the United States. For example, the organism infected 15 percent of AIDS patients in Haiti but only 0.2 percent of those in the United States.

Aquatic Bacterial Pathogens of Possible Concern for Potable Reuse Systems

Two types of aquatic microorganisms, aeromonads and cyanobacteria, may be of concern for potable reuse systems because their densities in water and/or their production of toxins could be influenced by wastewater nutrients. Indirect reuse systems that contain sufficient nutrients could create blooms of these organisms that may penetrate the treatment barriers and/or proliferate in the distribution system.

Aeromonads are commonly found in water and soil. Densities in water are related to fecal pollution and temperature, and aeromonads proliferate in domestic and industrial wastewater (Schubert, 1991). Some evidence suggests that *Aeromonas* may produce enterotoxins (Mascher et al., 1988), and several reports have suggested an association between gastroenteritis and *Aeromonas* in drinking water (Burke et al., 1984; Schubert, 1991). One study in Iowa concluded that three strains of *Aeromonas* were capable of causing diarrhea and that consumption of untreated water was a risk factor for *Aeromonas* infection (Moyer, 1987). A study in London found a correlation between *Aeromonas* isolates from water and isolates from fecal specimens (Nazer et al., 1990). However, two other studies reported little similarity between aeromonads isolated from diarrheal feces and those found in drinking water (Havelaar et al., 1992; Millership et al., 1988). Concern in the Netherlands about the possible health significance of aeromonads has led to the development of drinking water guidelines of less than 20 colony forming units (CFU) per 100 ml for drinking water leaving the treatment plant and less than 200 CFU/100 ml for drinking water in the distribution system (van der Kooij, 1993). Cooper and Danielson (1996) describe several methods of detecting these organisms in water and wastewater.

Cyanobacteria (blue-green algae) occur naturally in fresh and brackish waters worldwide. Although these are not infectious agents, some species produce toxins during algal blooms that are triggered by nutrient enrichment from natural waters, agricultural fertilizer runoff, or domestic or industrial effluents (Codd et al., 1989). Acute effects of exposure to these toxins have been well documented (Codd et al., 1989; Turner et al., 1990). Potential chronic health effects of long-term exposure to cyanobacterial toxins in drinking water are unknown. One study suggested that high rates of liver cancer in parts of China may be linked to cyanobacterial hepatotoxins in drinking water (Carmichael, 1994).

Control of cyanobacteria is problematic. Several studies indicate that the toxins can remain potent for days after the organisms have been destroyed by copper sulfate or chlorination (El Saadi et al., 1995; Hawkins et al., 1985). Using toxicity data from mouse bioassays, the Engineering and Water Supply Department of South Australia developed interim guidelines for acceptable numbers of cyanobacteria in water supplies (El Saadi et al., 1995). However, further research is needed on the acute and chronic toxicity of cyanobacterial toxins, and suitable methods need to be developed for monitoring the types and concentrations of cyanobacterial toxins in natural and treated water supplies (Elder et al., 1993).

OCCURRENCE OF MICROBIAL CONTAMINANTS IN MUNICIPAL WASTEWATER AND AMBIENT WATER

Disease epidemics or outbreaks are dramatic events, and our capability to identify the causes of such outbreaks is improving (described in Chapter 4). More difficult is determining whether microbial contaminants might cause occasional illnesses or low levels of disease (Craun et al., 1996; Frost et al., 1996). To assess this danger, researchers must rely on information regarding the pathogens' occurrences and concentrations in drinking water and extrapolate using the best data available on the pathogens' health risks.

The occurrence and concentration of pathogenic microorganisms in raw municipal wastewater depend on a number of factors that are not entirely predictable. Important variables include the source and original use of the water, the general health of the population, the existence of "disease carriers" for particular infectious agents, excretion rates of infectious agents, duration of the infection, and the ability of infectious agents to survive outside their hosts under various environmental conditions.

Untreated Wastewater

Fecal coliform bacteria, which are used as an indicator of microbial pathogens in wastewater, are typically found at 10^5 to 10^7 CFU per 100 ml in untreated wastewater (see Table 3-5). Other indicator bacteria, such as *Enterococci*, can range as high as 5 x 10^5 CFU per 100 ml (Rose et al., 1996). The bacterial pathogens of concern are usually found at much lower concentrations and, within the United States, typically range from 10^1 to 10^2 CFU per 100 ml in untreated wastewater. Levels of viruses in sewage vary greatly and reflect the variations in infection levels in the population, the season of the year (outbreaks of viral disease are often seasonal), and the methods used for their recovery and detection. Enteroviruses tend to be prevalent in the spring, and rotaviruses are more common in the winter (Gerba et al., 1985, 1996). In untreated domestic wastewater in the United States, virus counts typically range between 1000 and 71,000 plaque-forming units (PFU) per 100 liters (Danielson et al., 1996; Rose et

TABLE 3-5 Microorganism Concentrations in Untreated Municipal Wastewater

Microorganisms	Concentration (number per 100 ml)
Fecal coliforms	10^5-10^7
Fecal streptococci	10^4-10^6
Shigella	1-10^3
Salmonella	10^2-10^4
Pseudomonas aeruginosa	10^3-10^4
Clostridium perfringens	10^3-10^5
Helminth ova	1-10^3
Giardia lamblia cysts	10-10^4
Cryptosporidium oocysts	10^2-10^5
Entamoeba histolytica cysts	10^2-10^5
Enteric viruses	10^3-10^4

SOURCE: Adapted from National Research Council, 1996.

TABLE 3-6   Reported Levels of Pathogenic and Indicator Microorganisms in Untreated Wastewater

Reference	Average Levels Reported (CFU, PFU, or cysts/ oocysts per 100 liters)		
	Clostridium	Total Coliforms	Fecal Coliforms
Occoquan, Virginia (Rose et al., 1997)	36,667	2.4×10^7	9×10^5
St. Petersburg, Florida (Rose et al., 1996)	NT	8.2×10^7	2.2×10^7
South Africa (Grabow et al., 1978)	NT	2.46×10^5	NT
Tampa, Florida (City of Tampa, 1990)	NT	NT	NT
California (Yanko, 1993)[a]	NT	NT	NT
San Diego (Danielson et al., 1996)	NT	NT	NT
Denver (Lauer, 1991)	NT	8×10^5	4×10^4

NOTES: Only one study reported on levels of a bacterial pathogen: 2.2 CFU-MPN/ 100 ml of *Salmonella* (or 2200/100 liters) (Danielson et al., 1996). MPN = most probable number; NT = not tested.

al., 1996, 1997; Yanko, 1993). In South Africa they have been recorded as high as 4.0×10^5 viral units per 100 liters (Grabow et al., 1978).

Very few studies have examined the occurrence of enteric protozoa in wastewater. *Cryptosporidium* levels in untreated wastewater vary throughout the year from 100 to 1500 oocysts per 100 liters and are usually lower than enterovirus levels. *Giardia* is present in sewage at levels comparable to enteroviruses, averaging between 3,900 and 49,000 cysts/ 100 liters (Danielson, 1996; Rose et al., 1996, 1997). Table 3-6 summarizes the information available from microbiological studies of untreated wastewater. The more recent studies also monitored alternative microbiological indicator species—coliphage and *Clostridium*. The use of these indicators for evaluating potable reuse systems is reviewed in Chapter 4.

Primary and Secondary Effluent

Primary treatment does little to remove biological contaminants from

Enterococci	Coliphage	Enterovirus	Cryptosporidium Oocysts	Giardia Cysts
5×10^5	3.8×10^5	1,085	1,484	4.9×10^4
NT	2.8×10^6	1,000	1,500	6.9×10^3
6,400	3.8×10^4	71,000	NT	NT
NT	NT	7,000	30	3,900
NT	NT	5×10^3 4×10^4	NT	NT
NT	NT	2×10^3	2×10^2	3.25×10^4
Fecal streptococci[b] 7,000	5×10^4	NT	100	200

[a]Data reported from three reclamation plants.
[b]All fecal streptococci measured.

wastewater. However, some protozoa and parasite ova and cysts will settle out during primary treatment, and some particulate-associated microorganisms may be removed with settleable matter.

Secondary treatment, however, is designed to remove soluble and colloidal biodegradable organic matter and suspended solids. In some cases it also removes nitrogen and phosphorus. Secondary treatment consists of an aerobic biological process whereby microorganisms oxidize organic matter in the wastewater. The aerobic biological processes include activated sludge, trickling filters, rotating biological contactors, and stabilization ponds. Generally, primary treatment precedes these biological processes; however, some secondary processes, such as stabilization ponds and aerated lagoons, are designed to operate without sedimentation. Table 3-7 lists typical microorganism removal efficiencies for activated sludge and trickling-filter secondary treatment processes.

Conventional secondary treatment reduces pathogens but does not eliminate them from the effluent, even with disinfection. A Florida sur-

TABLE 3-7 Typical Percentage Removal of Microorganisms by Conventional Treatment Processes

| Microorganism | Primary Treatment | Secondary Treatment | |
		Activated Sludge	Trickling Filter
Fecal coliforms	<10	0-99	85-99
Salmonella	0-15	70-99	85-99+
Mycobacterium tuberculosis	40-60	5-90	65-99
Shigella	15	80-90	85-99
Entamoeba histolytica	0-50	Limited	Limited
Helminth ova	50-98	Limited	60-75
Enteric viruses	Limited	75-99	0-85

SOURCE: Reprinted, with permission, from Crook, 1992. © 1992 by Academic Press, Inc.

vey of wastewater treatment plants using activated-sludge secondary treatment after disinfection found viruses averaging 10 to 130 PFU/100 liters in 40 to 100 percent of the samples (Rose and Gerba, 1990). In a similar survey in California, 67 percent of the samples taken from secondary wastewater treatment facilities following disinfection contained viruses at levels ranging from 2 to 200 PFU/100 liters (Asano et al., 1992). Other studies of secondary effluent report similar findings, ranging from 3.5 to 650 PFU/100 liters (Rose et al., 1996, 1997; Yanko, 1993). However, Irving (1982) reported levels of enteroviruses as high as 715,000 viral PFU/100 liters. Likewise, protozoa can survive secondary treatment and disinfection. *Cryptosporidium* oocysts have been reported in secondary effluent at a level of 140 oocysts/100 liters (Rose et al., 1996), while *Giardia* cysts were found to range from 440 to 2297 cysts/100 liters (Rose et al., 1996, 1997). Table 3-8 summarizes the reported levels of pathogenic and indicator microorganisms in secondary effluent. These data suggest that wastewater discharges are contributing enteric pathogens to ambient waters, many of which may be used downstream for drinking purposes. All planned potable reuse projects and demonstration studies in the United States have used treatment in addition to secondary treatment, and such additional treatment is essential for protecting against risks of microbiological contamination.

Ambient Waters

In indirect reuse (either planned or unplanned), reclaimed water is discharged to a natural system (surface water or ground water), where it typically spends a period of time before being further treated for use as drinking water. During this time, natural processes tend to reduce the concentrations of enteric microorganisms beyond what occurs due to dilution alone. This "natural inactivation" or die-off rate is usually reported in terms of the time required for a 90 percent reduction in the viability of the microbial population. Many factors influence the inactivation rate, including the amount of solids, oxygen, salinity, and ultraviolet light the water is exposed to. However, temperature appears to play the most significant role. Enteric pathogens generally survive longer at lower temperatures. Table 3-9 summarizes the survival rates of some pathogens at selected temperatures.

The survival of the pathogenic bacteria corresponds closely to survival of the coliform indicator bacteria (Feachem et al., 1983; Korhonen and Martikainen, 1991; Olson, 1993; Singh and McFeters, 1990; Terzieve and McFeters, 1991). As shown in Table 3-9, the time required to achieve 90 percent reductions of *Salmonella* may range from 1 to 8 days at 10 to 20°C. At lower temperatures, between 5 and 10°C, a 99.9 percent reduction in the bacterial levels might require up to 24 days.

Virus survival is also related to water temperature. At ambient temperatures between 15 and 25°C, an inactivation rate of 99.9 percent can occur within 6 to 10 days. However, at 4°C, only 90 percent inactivation may occur after 30 days (Kutz and Gerba, 1988). The presence of undissolved solids may also aid virus survival. Kutz and Gerba (1988) showed that viruses survived longer in both sewage-polluted waters and ambient surface waters at any given temperature compared to tap water.

Very few data exist on survival of protozoan pathogens in ambient waters. DeRegnier et al. (1989) found that mice could no longer be infected with *Giardia* cysts collected after 56 days in river and lake water at 5°C. Using a more sensitive test (vital dyes) that can determine the potential viability of a single cyst, Robert et al. (1992) demonstrated that in river water between temperatures of 5 and 10°C, only 55 percent of the *Cryptosporidium* oocysts were dead after 47 days, 75 percent were dead after 60 days, and a 99 percent reduction in viability required 176 days. One can estimate a –0.01 to 0.05 $\log_{10}$ per day inactivation rate at low temperatures from their data.

MICROBIAL DATA FROM WATER REUSE APPLICATIONS

The public health hazards posed by microbial pathogens have been recognized since the practice of water reclamation and reuse began. Besides bacterial pathogens, viruses were a major concern, and almost all of the reuse projects and studies undertaken, whether pilot scale or full

TABLE 3-8 Reported Levels of Pathogenic and Indicator Microorganisms in Secondary Effluent From Wastewater Treatment Plants

Reference	Average Levels Reported (CFU, PFU, or cysts/oocysts per 100 liters)							
	Clostridium	Total Coliforms	Fecal coliforms	Enterococci	Coliphage	Enterovirus	Cryptosporidium oocysts	Giardia cysts
Occoquan, Virginia (Rose et al. 1997)	4,452	170,000	7,764	2,186	1,821	75.8	Not detected	2,297
St. Petersburg, Florida (Rose et al., 1996)	Not tested	1.5×10^6	190,000	Not tested	5×10^5	20	140	440
Tampa, Florida[a] (City of Tampa, 1990)	Not tested	Not tested	Not tested	Not tested	Not tested	3.5	Not tested	5
California (Yanko, 1993)[b]	Not tested	Not tested	Not tested	Not tested	Not tested	650 55 56	Not tested	Not tested

[a]Denitrified secondary effluent.
[b]Data reported from three reclamation plants.

TABLE 3-9 Survival of Enteric Pathogens and Indicator Bacteria in Fresh Waters

Microorganism	Time Reported (days) for 90 Percent Reduction in Viable Concentrations
Coliforms	0.83 to 4.8 days at 10 to 20°C, average 2.5 days
E. coli	3.7 days at 15°C
Salmonella	0.83 to 8.3 days at 10 to 20°C
Yersinia	7 days at 5-8.5°C
Giardia	14 to 143 days at 2 to 5°C
	3.4 to 7.7 days at 12 to 20°C
Enteric viruses	1.7 to 5.8 days at 4 to 30°C

SOURCES: Feachem et al., 1983; Korhonen and Martikainen, 1991; Kutz and Gerba, 1988; McFeters and Terzieva, 1991.

scale, began monitoring for enteric viruses in addition to the routine indicator bacteria.

Aside from water reuse projects, relatively few data exist regarding the levels of specific pathogenic microorganisms in wastewater or drinking water treatment processes, because monitoring is not routine or required in the United States. Neither federal nor state water quality standards specify concentrations of viruses or protozoa in drinking water, ambient waters, wastewaters, or reclaimed waters. Instead, microbial water quality standards have largely relied on bacterial indicators or treatment performance. Total coliform is used as a national standard for drinking water (the standard is less than 1/100 ml), while total or fecal coliforms are used in some states for reclaimed water. Indicators of treatment performance and water quality have been based on measurements of turbidity and suspended solids. More recently, enterococci, coliphage (a bacterial virus), and the Clostridium bacterium have been examined as biological indicators of treatment performance. (Chapter 4 further discusses microbial indicators.)

The increase in identified waterborne giardiasis and cryptosporidiosis outbreaks has made the drinking water industry more sensitive to protozoan contamination of water supplies. Through the Surface Water Treatment Rule (U.S. EPA, 1989), EPA developed performance standards for drinking water that require a 99.9 percent reduction of Giardia cysts and a 99.99 percent reduction of viruses by filtration and/or disinfection. EPA's goal is to achieve an annual risk no greater than a 1 in 10,000 chance of infection by a waterborne microbe from drinking water (Regli et al., 1991). While the rule does not specifically address wastewater contamination, it

TABLE 3-10   Concentrations of Parasites and Viruses in
Disinfected Secondary Effluents Used for Crop Irrigation in Arizona

Microbial Agent	Plant 1: Activated Sludge (aeration, chlorination)	Plant 2: Activated Sludge (aeration, denitrification, chlorination)
Giardia cysts/100 liters (positive samples)	66.8 (5/5)	1.57 (2/4)
Cryptosporidium oocysts/100 liters (positive samples)	(0/1)	No data
Enteroviruses PFU/100 liters (positive samples)	0.125 (3/52)	(0/54)

NOTES: Numbers in parentheses are number of positive samples per total samples
taken. Treated wastewater is being used for cotton crop irrigation. The Arizona
standard for public access irrigation was less than 2.5 PFU or cysts per 100 liters.

states that greater reductions may be required if a source water is of poor
quality. Due to the lack of monitoring information, EPA has recently
promulgated the Information Collection Rule, or ICR, to develop an occurrence database for *Cryptosporidium, Giardia*, and viruses in source waters, in some treated waters, and in various treatment processes. In light
of a national move toward watershed-based requirements, the ICR will
likely influence future microbial standards and monitoring requirements
pertaining to both planned potable reuse projects and potable water supplies influenced by upstream wastewater discharges.

Microbial Data From Nonpotable Reuse Applications

Arizona is currently the only state with standards and required monitoring for concentrations of viruses and *Giardia* in reclaimed water.
Arizona's current standard is less than or equal to 1 cyst or viral PFU per
40 liters (2.5 cysts or PFU/100 liters). The following sections summarize
the monitoring data available from Arizona, as well as the results of specific microbial studies from California and Florida based largely on ter-
tiary-treated wastewater (secondary treatment, filtration, and chlorination) used for nonpotable reuse applications.

Plant 3: Lagoon (5 days retention time, mechanical aeration, UV light disinfection)	Plant 4: Biotowers (compressed air, chlorination)	Plant 5: Activated Sludge (oxygen, chlorination)	Plant 6: Lagoon (mechanical aeration, chlorination)
18.3 (6/11)	26 (35/38)	43.5 (36/42)	(0/7)
11.4 (2/2)	3.4 (16/30)	3.7 (15/34)	1.5 (2/5)
7.75 (11/45)	0.725 (6/16)	0.75 (3/47)	No data

SOURCE: C. P. Gerba, personal communication, 1996.

Microbial Monitoring in Arizona

Data on concentrations of *Giardia, Cryptosporidium,* and enteroviruses are available from wastewater and reclamation facilities in Arizona where the effluent is used for irrigation. Arizona currently has no requirements for monitoring of *Cryptosporidium* in reclaimed waters; however, this protozoan was included in most monitoring programs. Monitoring frequency is established on a case-by-case basis and is determined partly by the flow, treatment design, and designated reuse application. Frequency ranges from once per month to twice per year. Table 3-10 summarizes the monitoring results for six reclamation facilities that use a variety of secondary treatment options followed by disinfection with chlorination or ultraviolet light. The effluents were used primarily for irrigating cotton crops.

Collectively, *Giardia* was found in 78.5 percent of the effluent samples from all plants at an average concentration of 31.3 cysts/100 liters. *Cryptosporidium* was found in 59 percent of the samples from all plants at an average concentration of 5 oocysts/100 liters. Viruses were found in 18 percent of the samples from all plants at an average concentration of 2.2 most probable number (MPN) PFU per 100 liters. No differences in protozoa levels were readily detected in the two plants using lagoon effluents.

Table 3-11 illustrates the efficacy of combining sand filtration and

TABLE 3-11 Concentrations of Parasites and Viruses in Filtered, Disinfected Secondary Effluents in Arizona

Microbial Agent	Plant 7: Lagoon (aeration, filtration, chlorination)	Plant 8: Activated Sludge (aeration, filtration, chlorination)	Plant 9[a] : Filtration (deep bed dual media sand and coal pressure filters, chlorination)
Giardia cysts/100 liters (positive samples)	11.25 (10/16)	7.4 (4/9)	6.98 (25/50)
Cryptosporidium oocysts/100 liters (positive samples)	No data	1.88 (1/2)	3.02 (17/50)
Enteroviruses PFU/100 liters (positive samples)	No data	No data	0.15 (2/45)

NOTE: Numbers in parentheses are number of positive samples per total samples taken. Arizona's standard for public access irrigation was less than 2.5 PFU or cysts per 100 liters.

[a]Plant 9 operated with secondary treated sewage from Plant 4 in Table 3-10.

SOURCE: C. P. Gerba, personal communication, 1996.

disinfection after secondary treatment. The effluents from these three plants are used to irrigate golf courses. Giardia was found in 55 percent of all the samples from the three plants at an average concentration of 10 cysts/100 liters. This represents a reduction of 70 percent compared to the nonfiltered effluents in Table 3-10. Cryptosporidium oocysts were detected in 56.2 percent of all samples at an average concentration of 2.5 oocysts/100 liters, representing a reduction of 51 percent compared to nonfiltered effluent.

The filter plants varied in design; however, none of the plants used coagulants during the filtration process, which would have further improved protozoa removal. Viruses were detected in only 4.4 percent of the samples in Plant 9 (the only plant that sampled for viruses) at a level of 0.15 MPN-PFU/100 liters. This represents a reduction of 93.5 percent compared to nonfiltered effluents. Dual-media filtration is particularly effective in removing suspended solids and turbidity, which enhances the efficacy of chlorination.

TABLE 3-12 Occurrence of Microorganisms Throughout Treatment Train at St. Petersburg, Florida, Reclamation Facility

Parameter	Untreated Wastewater	Post-clarification	Post-filtration	Post-chlorination	Storage Tank
Total Coliforms					
Percent positive	100	100	100	18	18
Concentration (CFU/100 ml)	$8.2 \pm 2.3 \times 10^7$	$1.5 \pm 1.6 \times 10^6$	$4.6 \pm 4.8 \times 10^5$	$2.7 \pm 4.8 \times 10^1$	6.6 ± 1.1
Fecal Coliforms					
Percent positive	100	100	100	9	9
Concentration (CFU/100 ml)	$2.2 \pm 0.6 \times 10^7$	$1.9 \pm 0.2 \times 10^5$	$1.7 \pm 1.5 \times 10^5$	1.9 ± 2.3	0.82 ± 1.2
Phage					
Percent positive	100	90	92	50	25
Concentration (PFU/100 ml)	$2.8 \pm 9.8 \times 10^6$	$5.0 \pm 9.0 \times 10^5$	$7.6 \pm 7.6 \times 10^1$	7.2 ± 6.0	0.70 ± 1.1
Enteroviruses					
Percent positive	100	58	50	25	8
Concentration (PFU/100 liters)	$1.0 \pm 1.4 \times 10^3$	$2.0 \pm 2.1 \times 10^1$	3.1 ± 3.2	0.33 ± 0.13	0.01 ± 0.01
Giardia					
Percent positive	100	83	75	42	25
Concentration (cysts/100 liters)	$6.9 \pm 3.7 \times 10^3$	$4.4 \pm 4.7 \times 10^2$	4.4 ± 3.7	0.97 ± 1.5	0.49 ± 1.5
Cryptosporidium					
Percent positive	67	42	42	25	17
Concentration (oocysts/100 liters)	$1.5 \pm 1.8 \times 10^3$	$1.4 \pm 1.7 \times 10^2$	4.0 ± 2.2	1.8 ± 1.5	0.75 ± 1.1

SOURCE: Reprinted, with permission, from Rose et al., 1996. ©1996 by Elsevier Science.

TABLE 3-13 Reduction of Microorganisms by Process for the St. Petersburg, Florida, Reclamation Facility

	Unit Process Reduction ($\log_{10}$)				
Parameter	Biological Treatment and Clarification	Filtration	Chlorination	Storage	Complete Treatment
Total coliforms	1.75	0.51	4.23	0.61	7.10
Fecal coliforms	2.06	0.05	4.95	0.36	7.42
Phage	0.75	3.81	1.03	1.03	6.62
Enterovirus	1.71	0.81	1.45	1.04	5.01
Giardia	1.19	2.00	0.65	0.30	4.14
Cryptosporidium	1.14	1.68	0.41	0.04	3.27

SOURCE: Reprinted, with permission, from Rose et al., 1996. ©1996 by Elsevier Science.

Virus Studies in California

In the 1970s, California adopted stringent treatment requirements for reclaimed waters destined for public access applications. The treatment was designed to reduce viruses by 5 $\log_{10}$ (99.999 percent) and included coagulation, flocculation, sedimentation, sand filtration, and disinfection with a 5 mg/liter total chlorine residual for 1.5 hours. Asano et al. (1992) summarized the virus monitoring data for Orange County Sanitation District, the Monterey Regional Water Pollution Control Agency, the County Sanitation Districts of Los Angeles, and the Las Virgines Municipal Water District—a 3- to 10-year data set. In the unchlorinated secondary effluents, 66.7 percent of 424 samples collected were positive for viruses. Geometric averages ranged from 2 to 200 MPN-PFU/100 liters. In the finished product water, Asano et al. (1992) found less than 1 percent of 814 samples positive for viruses. These few positives were attributed to operational difficulties during chlorination. Yanko (1993) found similar results in summarizing 10 years of analysis of six reclamation facilities in Los Angeles County. He reported that only 0.17 percent (1/590) of the samples from reclamation facilities were positive for viruses.

Microbial Studies in Florida

Florida requires less stringent treatment than California for reclaimed water that may be used in areas accessible to the public. Performance criteria specify secondary treatment plus sand filtration to meet a suspended solids standard and a minimum chlorination level of 1 mg/liter for 15 minutes of peak flow to meet a fecal coliform standard of less than 1/100 ml. St. Petersburg has been operating a full-scale wastewater treatment and reclamation facility for more than 20 years. This plant produces 16 million gallons of reclaimed water per day. Advanced treatment includes dual-media rapid sand filtration with in-line addition of alum and polymer coagulants. The final effluent is chlorinated and stored in an 8 million gallon tank for an average of 16 to 24 hours prior to release into a distribution system used for golf course and residential landscape irrigation. Rose et al. (1996) evaluated the St. Petersburg treatment processes and documented partial removal of bacteriophages and enteroviruses at each stage of treatment (see Tables 3-12 and 3-13). Bacteriophages were reduced by the total treatment process by 6 $\log_{10}$, and enteroviruses were reduced by 5 $\log_{10}$. An average of 0.7 PFU/100 liters of bacteriophages was found in 25 percent of the samples from the storage tank containing the final reclaimed water. This facility maintains a much more stringent disinfection standard (4 mg/liter chlorine and 45 minutes contact time) than Florida requires, and the virus reductions are similar to those seen in California facilities.

St. Petersburg's advanced treatment process reduced the numbers of *Giardia* cysts and *Cryptosporidium* oocysts by 4.14 and 3.27 $\log_{10}$, respectively, and removal was observed at each stage of treatment (Rose et al., 1996). In the final effluent from the storage tank, 25 percent of the samples were still positive for *Giardia* cysts and 17 percent were positive for *Cryptosporidium* oocysts. Methodological limitations prevented a determination of the viability of the cysts and oocysts detected.

In a study of drinking water treatment, researchers reported a 99 percent inactivation rate for *Giardia* cysts after 50 to 180 minutes of contact time with 5 to 16 mg/liter of monochloramine (Hoff, 1986). Studies of *Cryptosporidium* in drinking water suggest that levels of 15 mg/liter of chloramines for 240 minutes are needed to reduce viable oocysts by 99 percent (Finch, 1994). In the St. Petersburg reclamation facility, the storage tank contains an average of 2.5 mg/liter of total chlorine and involves approximately 16 to 24 hours more contact time. No studies have examined the inactivation of protozoan cysts and oocysts in wastewater or reclaimed water, leaving the mechanisms for inactivation and the efficacy of chlorine disinfection against *Giardia* cysts or *Cryptosporidium* oocysts poorly understood.

Advanced Treatment for Potable Reuse

Table 3-14 summarizes microbial monitoring data available from seven potable reuse facilities using advanced treatment. Orange County Water District conducted two significant microbial studies at the Water

TABLE 3-14 Monitoring Data at Potable Reuse Projects for
Microbial Pathogens of Concern

Facility (barriers to pathogens)[a]	Testing Performed	Cultivable Enteric Viruses
Denver (lime, sand filtration, carbon, RO, or UF)	Concentration	<1/1000 liters
	Positive samples	0/37
	Percent reduction	n.a.
Water Factory 21 1979 study (lime, ammonia stripping, carbon, RO, chlorination)	*After Lime*	
	Positive samples	28/28
	Percent reduction	99.87%
	After Chlorination	
	Positive samples	1/142
Water Factory 21 1980-1981 study (lime, ammonia stripping, carbon, RO, chlorination)	Concentration	<0.1/100 liters
	Positive samples	0/21
	Percent reduction	>99%
	Prechlorination	
	Concentration	0.2/100 liters
	Positive samples	1/19
	Percent reduction	99.4%
Potomac study (lime, intermediate chlorination, dual-media filtration, carbon, chlorination or ozonation)	Concentration	<1/1700 liters
	Positive samples	0/56
	Percent reduction	>87%
Tampa (lime; sand filtration; RO, UF, or carbon; ozonation or chlorination)	*After Lime*	
	Concentration	0.06/100 liters
	Positive samples	4/25
	Percent reduction	98.3%
	After Chlorination	
	Concentration	0.02/100 liters
	Positive samples	1/15
	Percent reduction	99.4%
	After Ozone	
	Concentration	<0.01/100 liters
	Positive samples	0/4
	Percent reduction	>99%

Cryptosporidium	Giardia	Other Microbial Pathogens Examined
<1/100 liters 0/4 >97.5%	<1/100 liters 0/15 >99.4%	*Shigella, Salmonella, Campylobacter, Entamoeba* tested, none detected
Not tested	Not tested	Not tested
Not tested	Not reported 0/20 0% (+); after chlorination and RO <0.05/100 liters — >86.9% removals	Helminths: none found
Not tested	None detected using light microscopy	Not tested
0.13 /100 liters 1/16 99.6%	<1/200 liters in final effluent >99.97%	
<1/200 liters[b] 0/6 >99.8%		

Table continues on next page

TABLE 3-14 Continued

Facility (barriers to pathogens)[a]	Testing Performed	Cultivable Enteric Viruses
San Diego (Aqua III, water hyacinth, ponds, dual-media filtration, UV, RO, carbon)	Concentration Positive samples Percent reduction	<1/1000 liters 0/32 >99.995%
Windhoek, South Africa (ponding, lime, sand filtration, chlorination, carbon)	Concentration Positive samples Percent reduction	1/10 liters 0/31 n.a.
Upper Occoquan Sewage Authority (lime, filtration, carbon, chlorination)	Concentration Positive samples Percent reduction	<1/500 liters 0/11 >99.995%[c]

[a]All testing is on final effluent unless otherwise noted. Carbon = carbon adsorption; lime = chemical lime treatment, pH 11.2, recarbonation; RO = reverse osmosis; UF = ultrafiltration; UV= ultraviolet disinfection.
[b]After filtration.
[c]Based on raw sewage counts; all other removals based on counts entering reclamation facility.

Factory 21 reclamation plant, in 1979 and 1981. In the 1979 study (James M. Montgomery, Inc., 1979), one of the 142 final, disinfected effluent samples was positive for viruses, and the study recommended that the disinfection procedure be optimized to increase the contact time and to increase chlorine residual to 5 mg/liter. In the second study (James M. Montgomery, Inc., 1981), no samples were positive for viruses in the final effluent, an improvement due to improved disinfection and the reduction of influent virus levels by upgrading the secondary treatment from a trickling filter effluent to an activated sludge system. This study also demonstrated that reverse osmosis was not a good substitute for disinfection. Without disinfection, 5.3 percent of the samples from reverse osmosis effluent still contained viruses. Studies conducted by San Diego during the Aqua II project also noted relatively high virus breakthrough in reverse osmosis systems (Western Consortium for Public Health, 1992). During the Tampa project, viruses were detected in 6.7 percent of the samples after chlorination, but this occurred during an operational period when pH levels were suboptimal during lime treatment and there was a loss in normal chlorine residuals (Western Consortium for Public Health, 1992).

Cryptosporidium	Giardia	Other Microbial Pathogens Examined
1/1000 liters	<1/1000 liters	1/500 ml (*Salmonella*)
2/29[d]	0/29	0/29
99.995%	>99.9997%	>99%
Not tested	Not tested	Not tested
0.44 /100 liters	6.6 /100 liters	
1/11	2/11	
99.97%	99.986%	

[d]Positives seen during spiking trials.

SOURCES: CH$_2$M Hill, 1993; Grabow, 1990; James M. Montgomery, 1979, 1981, 1983; Lauer et al., 1990; Western Consortium for Public Health, 1992.

At two reclamation facilities in San Diego and Denver, seeding studies were conducted in which viruses were artificially inoculated in the secondary influent feedwater to advanced wastewater treatment. The results suggest that as much as a 10 $\log_{10}$ reduction of viruses can be achieved by multiple barriers within a reclamation facility (Lauer et al., 1991; Western Consortium for Public Health, 1992). Therefore, the main issues for virus control are the known levels entering the plant, the numbers of barriers required, and the reliability of those barriers.

The San Diego Aqua II project evaluated protozoa and detected no *Giardia* cysts in 29 samples from the final effluent; however, *Cryptosporidium* oocysts were detected after a spiking trial (Western Consortium for Public Health, 1992). This suggests that if high concentrations of oocysts were to enter the facility, a few might penetrate the barriers. The San Diego study also demonstrated that protozoa can be detected at a limit of 1 cyst/oocyst in 1000-liter samples of highly treated reclaimed water. The Tampa study (CH$_2$M Hill, 1993) examined individual unit processes and demonstrated that chemical lime treatment removed 99 percent of the cysts and oocysts. However, Rose et al. (1997) found no decrease in viability of the oocysts after exposure to high pH or to a

TABLE 3-15 Reductions of Pathogenic and Indicator Microorganisms in the UOSA Reclamation Facility Compared to Influent Concentrations

After Treatment in Reclamation Facility	Percentage and Log_{10} Reductions of Microorganisms Compared to Untreated Wastewater		
	Clostridium	Total Coliform	Fecal Coliform
Secondary treatment	−0.92	−2.15	−2.06
	87.8	99.29	99.14
Chemical lime treatment	−3.86	−5.3	−4.59
	99.986	99.9995	99.997
Poststabilization basin[a]	−3.35	−4.12	−3.2
	99.955	99.992	99.94
Multimedia filtration	−4.31	−5.13	−4.39
	99.995	99.9992	99.996
GAC upflow adsorption	−4.22	−5.16	−4.63
	99.994	99.9993	99.998
Chlorination	−4.02	−7.18	>−5.95
	99.9905	99.999993	>99.9999

NOTE: GAC = granular activated carbon.

[a]Influent to multimedia filtration through an open basin.

combination of high pH and disinfection in studies at the Upper Occoquan Sewage Authority (UOSA). Therefore the removal mechanism for protozoa after chemical lime treatment appears to be physical removal of the oocysts. Rose et al. (1997) also found that multimedia filtration further reduced enteric protozoa by 85 to 95.7 percent. However, protozoan cysts and oocysts were still detected in the final effluent of the UOSA facility, which incorporates lime, multimedia filtration, carbon absorption, and chlorination.

Approximately 4 to 6 log_{10} removals of *Giardia* and *Cryptosporidium* cysts and oocysts have been documented in both San Diego and UOSA by a combination of processes. Table 3-15 shows the percentage and log

Enterococci	Coliphage	Enterovirus	Cryptosporidium	Giardia
-2.36	-2.62	-1.16		-1.3
99.56	99.76	93		95.3
-4.33	-4.88	-3.69	-2.63	-2.46
99.995	99.998	99.98	99.8	99.65
-3.67	-4.82	-3.56	-1.67	-2.34
99.98	99.998	99.97	97.9	99.55
-3.61	-4.57	-3.48	-3.22	-2.90
99.97	99.997	99.97	99.94	99.87
-4.67	-4.62	-3.52	-2.74	
99.998	99.998	99.97	99.82	
-5.27	-5.86	>-4.34	-3.53	-3.87
99.9995	99.99986	>99.995	99.97	99.986

reduction of pathogens and microbial indicators by unit process in the UOSA facility (Rose et al., 1997). The bacterial indicator *Clostridium* best reflects the removal of enteroviruses for secondary treatment and chemical lime treatment. Coliphage appears to better reflect the removal of viruses during the disinfection process.

Reverse osmosis was found to be the single most effective barrier to cysts and oocysts. Chemical treatment was the next most effective and sand filtration the least. No studies to date have examined the disinfection of cysts and oocysts or the optimization of sand filtration in wastewater or reclaimed water.

TABLE 3-16 Relative Removal of Pathogens and Coliform Indicators by Various Treatment Processes

Unit Process	Enterovirus	*Giardia*	*Cryptosporidium*	Coliform
Biological secondary treatment	+	+	ND	+
Coagulation-flocculation-sedimentation-filtration	++	++	+	++
Chlorination (free)	+++	+	–	++++
Combined chlorine	+	–	–	++
Ozone disinfection	++++	++	+	++++
UV disinfection	++	+	ND	++++
Reverse osmosis	+++	+++	+++	+++
Lime treatment[a]	++	++	++	+++
Microfiltration	+	+++	++++	++++
Ultrafiltration	++++	++++	++++	++++

NOTE: + signs indicate removal from low (+) to very high (++++); minus sign indicates no significant removal; ND indicates absence of data to make a judgment. For disinfectants, assessment is based on the rate of inactivation of organisms rather than log removals.

[a]Chemical lime treatment adds a disinfection barrier to bacteria and viruses due to its high pH and a removal barrier to protozoa via the precipitate formed and the sedimentation process.

Chemical treatment, disinfection (chlorine or ultraviolet), and reverse osmosis are effective barriers for the removal of viruses. Effective barriers for removing protozoa include chemical treatment, reverse osmosis, and, to lesser degree, sand filtration. The efficacy of disinfection, in particular ozone, awaits further evaluation for protozoa removal. Table 3-16 summarizes the relative efficiency of various unit processes in water reclamation systems as barriers to microbial pathogens.

CONCLUSIONS

Microbial contaminants in reclaimed water include the enteric bacteria, enteric viruses, and enteric protozoan parasites. Classic waterborne bacterial diseases, such as dysentery, typhoid, and cholera, while still important worldwide, have dramatically decreased in the United States. However, *Campylobacter*, nontyphoid *Salmonella*, and pathogenic *Escherichia coli* still cause a significant number of illnesses, and new emerging diseases also pose potentially significant health risks.

Historically, coliforms have served as an effective indicator for many bacterial pathogens of concern. However, most recognized outbreaks of waterborne disease in the United States are caused by protozoan and viral pathogens in waters that have met current coliform standards. Table 3-17 summarizes how the three main microorganisms of concern, *Giardia, Cryptosporidium*, and the enteric viruses, rank with regard to their occurrence in wastewater, resistance to water treatment, adequacy of monitoring, and severity of health risk. *Giardia* is one of the most frequently identified microbial pathogens, occurring consistently in high numbers in untreated wastewater, secondary effluent, and secondary effluent receiving sand filtration and disinfection. However, the health threat it poses is relatively low because the resulting gastroenteritis is less severe and more amenable to treatment than infections caused by the viruses or *Cryptosporidium*. *Cryptosporidium*, which may cause severe diarrhea in immunocompromised individuals, is found in highly variable levels in wastewater. It is highly resistant to disinfection and difficult to detect in untreated wastewater with current methods. Studies in California show that disinfection standards using a concentration/contact time approach can reliably reduce enteroviruses in reclaimed waters. However, monitoring has not been conducted for other viruses of concern, such as adenoviruses, rotaviruses, and Norwalk and related human caliciviruses.

Wastewater may also contain a number of newly recognized or "emerging" waterborne enteric pathogens or potential pathogens. For some of these organisms there is no evidence of waterborne transmission, and their occurrence in wastewater is suspected but not documented.

RECOMMENDATIONS

To ensure the safety of drinking water produced from reclaimed water, planners, regulators, and operators of potable reuse systems must account for the various existing and potential health risks posed by microbial contaminants.

• **Potable reuse systems should continue to employ a combination of advanced physical treatment processes and strong chemical disinfectants as the principal line of defense against most microbial contaminants.** Some new membrane water filtration systems can almost completely remove microbial pathogens of all kinds, but experience with them is not yet adequate to depend on them alone for protection against the serious risks posed by these pathogens. Therefore, strong chemical disinfectants, such as ozone or free chlorine, should also be used, even in systems that include membrane filters.

• **Current and future facilities should assess and report the effectiveness of their treatment processes in removing microbial pathogens so that the industry and regulators can develop guidelines and stan-**

TABLE 3-17 Ranking of Microbial Contaminants of Most Concern in Reclaimed Water

Relative Rank	Concentration in Wastewater	Variation in Concentration
Most concern	*Giardia* at highest levels, always present	*Giardia* appears to be constant
Moderate concern	Enteroviruses, always present	Enteroviruses moderately variable
Least concern	*Cryptosporidium* sometimes present	*Cryptosporidium* highly variable, but often undetectable

dards for operations. Facilities should report number of barriers, microbial reduction performance, and the reliability or variation. They should conduct seeded tracer and pilot studies to provide data on performance in addition to information on the occurrence, concentrations, and variations in loadings of indigenous microorganisms. Appropriate disinfection studies should also be undertaken for the enteric protozoa and for some viruses.

• **To provide protection against emerging pathogens, the EPA should support research to develop methods for detecting emerging pathogens in environmental samples.** Research is also needed on the effectiveness of various water or wastewater treatment processes and disinfectants in removing or inactivating these pathogens.

• **Both the industry and the research community need to establish the performance and reliability of individual barriers to microorgan-**

Resistance to Treatment	Barriers Available to Control Microorganism	Ability to Monitor Microorganisms or Adequate Surrogate	Health Outcome
Cryptosporidium most resistant	*Giardia* by filtration and possibly disinfection (no data on wastewater or reclaimed water disinfection of cysts)	*Giardia* cysts directly	Some enteroviruses cause serious illness (e.g., heart disease, chronic sequelae)
Giardia more resistant to disinfection than bacteria or viruses	Enteroviruses by disinfection	Enteroviruses directly; coliphage is a possible surrogate	*Cryptosporidium* diarrhea; 1.0% hospitalization; 50% mortality in immuno-compromized population
Enteroviruses more resistant than bacteria	*Cryptosporidium* by filtration	*Cryptosporidium* oocysts directly; *Clostridium* is a possible surrogate	*Giardia* causes diarrhea, sometimes chronic; 0.45% hospitalization; 0.0001% mortality

isms within treatment trains and to develop performance goals appropriate to planned potable reuse. Most present regulations and guidelines for microbial water quality and treatment performance are based on nonpotable reuse studies focusing on incidental or recreational contact with reclaimed water. Since potable reuse poses greater risks, existing state reuse regulations may not be sufficiently stringent. And while national standards for water treatment are based on scientific risk assessment procedures, they generally assume that the source water is natural surface water or ground water. Potable reuse projects, as well as water sources that are heavily impacted by upstream wastewater discharges, may need to achieve greater levels of pathogen reduction. Only California has established treatment barriers for viruses that are more protective than those used in ordinary drinking water treatment facilities. Treatment standards and goals more appropriate to potable reuse projects need to be developed so they will be in place as the number of potable reuse projects increases.

REFERENCES

Abad, F. X., R. M. Pinto, C. Villena, R. Gajardo, and A. Bosch. 1997. Astrovirus survival in drinking water. Applied and Environmental Microbiology 63:3119-3122.

Asano, T., L. Y. C. Leong, M. G. Rigby, and R. H. Sakaji. 1992. Evaluation on the California wastewater reclamation criteria using enteric virus monitoring data. Water Science and Technology.

Bennett, J. V., S. D. Homberg, M. F. Rogers, and S. L. Solomon. 1987. Infectious and Parasitic diseases. American Preventive Medicine 3:102-114.

Bryan, R. T., A. Cali, R. L. Owen, and H. C. Spencer. 1991. Microsporidia: Opportunistic pathogens in patients with AIDS. Pp. 1-26 in T. Sun (ed) Progress in Clinical Parasitology. Philadelphia: Field and Wood.

Burke, V., J. Robinson, M. Gracey, D. Peterson, and K. Partridge. 1984. Isolation of *Aeromonas hydrophila* from a metropolitan water supply: seasonal correlation with clinical isolates. Applied and Environmental Microbiology 48:361-366.

Carmichael, W. W. 1994. The toxins of cyanobacteria. Scientific American :78-86.

Caul, E. O. 1994. Human coronaviruses. Pp. 603-625 in Kapikian, A. Z. (ed.) Viral Infections of the Gastrointestinal Tract. New York: Marcel Dekker.

Centers for Disease Control and Prevention (CDC). 1993. Surveillance for waterborne disease outbreaks—United States, 1991-1992. MMWR 42:1-22.

Centers for Disease Control and Prevention (CDC). 1990. Waterborne disease outbreaks. U.S. Department of Health and Human Services, Atlanta, Ga. Morbidity and Mortality Weekly Report 39(SS-1):1-57

CH_2M Hill. 1993. Tampa Water Resources Recovery Project Pilot Studies, Volume 1 Final Report. Tampa, Fla.: CH_2M Hill.

Codd, G. A., S. G. Bell, and W. P. Brooks. 1989. Cyanobacterial toxins in water. Water Science and Technology 21:1-13.

Cooper, R. C., and R. E. Danielson. 1996. Detection of bacterial pathogens in wastewater and sludge. Pp. 222-230 in C. J. Hurst et al. (eds.) Manual of Environmental Microbiology. Washington, D.C.: ASM Press.

Crabtree, K. D., C. P. Gerba, J. B. Rose, and C. N. Haas. 1997. Waterborne adenovirus: A risk assessment. Water Science and Technology 11-12: 1-6.

Craun, G. F. 1986. Waterborne Diseases in the United States. Boca Raton, Fla.: CRC Press.

Craun, G. F. 1991. Statistics of waterborne disease in the United States. Water Science and Technology 24(2):10-15.

Craun, G. F., R. L. Calderon, F. J. Frost. 1996. An introduction to epidemiology. Journal of the American Water Works Association 88(9):54-65.

Crook, J. 1992. Water Reclamation. Pp. 559-589 In Encyclopedia Physical Science and Technology, Vol. 17. San Diego, Calif.: Academic Press.

Current, W. L., and L. S. Garcia. 1991. Cryptosporidiosis. Clinical Microbiology Review 4(3):325-358.

Danielson, R. E., L. A. Pettegrew, J. A. Soller, A. W. Olivieri, D. M. Eisenberg, and R. C. Cooper. 1996. A microbiological comparison of a drinking water supply and reclaimed wastewater for direct potable reuse. Paper presented at the Joint AWWA and WEF Water Reuse 96 Conference held in San Diego, Calif., January 1996.

DeRegnier, D., L. Cole, D. G. Schupp, and S. L. Erlandson. 1989. Viability of *Giardia* cysts suspended in lake, river and tap water. Applied and Environmental Microbiology 55(5):1223-1229.

Enroth, H., and L. Engstrand. 1995. Immunomagnetic separation and PCR for detection of *Helicobacter pylori* in water and stool samples. Journal of Clinical Microbiology 33:2162-2165.

El Saadi, O., A. J. Esterman, S. Cameron, and D. M. Roder. 1995. Murray River water, raised cyanobacterial cell counts, and gastrointestinal and dermatological symptoms. Medical Journal of Australia 162:122-125.

Elder, G. H., P. R. Hunter, and G. A. Codd. 1993. Hazardous freshwater cyanobacteria (blue-green algae). Lancet 341:1519-1520.

Fayer, R. G., and B. L. P. Ungar. 1986. *Cryptosporidium* spp. and cryptosporidiosis. Microbiology Review 50(4): 458-483.

Feachem, R. G., D. J. Bradley, H. Garelick, and D. D. Mara (eds.). 1983. *Entamoeba histolytica* and amebiasis. Pp. 337-347 in Sanitation and Disease: Health Aspects of Excreta and Wastewater Management. New York: John Wiley and Sons.

Fiedorek, S. C., H. M. Malaty, D. L. Evans, C. L. Pumphrey, H. B. Casteel, D. J. Evans, and D. Y. Graham. 1991. Factors influencing the epidemiology of *Helicobacter pylori* infection in children. Pediatrics 88:578-582.

Finch, G. R., B. Kathleen, and L L. Gyurek. 1994. Ozone and chlorine inactivation of *Cryptosporidium*. Pp. 1303-1320 in Proceedings of the American Water Works Association's Water Quality Technology Conference, San Francisco, Calif., November 6-10. Denver, Colo.: American Water Works Association.

Foy, H. M. 1991. Adenoviruses. Pp. 77-94 in Evans, A. S. (ed.) Viral Infections of Humans: Epidemiology and Control. New York: Plenum Medical Book Company.

Frost, F. J., G. F. Craun, and R. L. Calderon. 1996. Waterborne disease surveillance. Journal of the American Water Works Association 88(9): 66-75.

Geldreich, E. E., K. R. Fox, J. A. Goodrich, E. W. Rice, R. M. Clark, and D. L. Swerdlow. 1992. Searching for a water supply connection in the Cabool, Missouri, disease outbreak of *Escherichia coli* O157:H7. Water Research 26:1127-1137.

Gerba, C. P., S. N. Singh, and J. B. Rose. 1985. Waterborne viral gastroenteritis and hepatitis. CRC Crit. Rev. in Environmental Control 15:213-236.

Gerba, C. P., and J. B. Rose. 1993. Comparative environmental risk. Pp. 117-135 in Cothern, C. R. (ed.) Estimating Viral Disease Risk From Drinking Water. Boca Raton, Fla.: Lewis Publishers.

Gerba, C. P., J. B. Rose, and C. N. Haas. 1994. Waterborne disease: Who is at risk? in Proceedings of the American Water Works Association's Water Quality Technology Conference, San Francisco. Denver, Colo.: American Water Works Association.

Gerba. C. P., J. B. Rose, and C. N. Haas. 1996. Sensitive population: Who is the greatest risk? International Journal of Food Microbiology 30:113-123.

Goodgame, R. W. 1996. Understanding the intestinal spore-forming protozoa: *Cryptosporidium*, microsporidia, *Isospora* and *Cyclospora*. Annals of Internal Medicine 124:429-441.

Grabow, W. O. K., B. W. Bateman, and J. S. Burger. 1978. Microbiological quality indicators for routine monitoring of wastewater reclamation systems. Prog. Wat. Tech 10:317-327.

Grabow, W. O. K. 1990. Microbiology of drinking water treatment: Reclaimed wastewater. Pp. 185-203 in McFeters, G. A. (ed.) Drinking Water Microbiology: Progress and Recent Developments. New York: Springer-Verlag.

Hale, D., W. Aldeen, and K. Carroll. 1994. Diarrhea associated with cyanobacterialike bodies in an immunocompetent host. JAMA 271:144-145.

Havelaar, A. H., F. M. Schets, A. van Silfhout, W. H. Jansen, G. Wieten, and D. van der Kooij. 1992. Typing of *Aeromonas* strains from patients with diarrhoea and from drinking water. Journal of Applied Bacteriology 72:435-444.

Hawkins, P. R., M. T. C. Runnegar, A. R. B. Jackson, and I. Falconer. 1985. Severe hepatotoxicity caused by the tropical cyanobacterium (blue-green alga) *Cylindrospermopsis raciborskii* (Woloszynska) Seenaya and Subba Raju isolated from a domestic water supply reservoir. Applied and Environmental Microbiology 50:1292-1295.

Herwaldt, B. L., G. F. Craun, S. L. Stokes, and D. D. Juranek. 1992. Outbreaks of waterborne disease in the United States: 1989-90. Journal of the American Water Works Association 84:129-135.

Hoff, J. C. 1986. Inactivation of Microbial Agents by Chemical Disinfectants. EPA/600/2-861067. Washington, D.C.: EPA, Drinking Water Research Division.

Hopkins, R. J., P. A. Vial, C. Ferreccio, J. Ovalle, P. Prado, V. Sotomayor, R. G. Russell, S. S. Wasserman, and J. G. Morris. 1993. Seroprevalence of *Helicobacter pylori* in Chile: Vegetables may serve as one route of transmission. Journal of Infectious Diseases 168:222-226.

Huang P., J. T. Weber, D. M. Sosin, P. M. Griffin, E. G. Long, J. J. Murphy, F. Kocka, C. Peters, and C. Kallick. 1995. The first reported outbreak of diarrheal illness associated with *Cyclospora* in the United States. Annals of of Internal Medicine 123:409-414.

Hurst, C. J., W. H. Benton, and R. E. Stetler. 1989. Detecting viruses in water. Journal of the American Water Works Association 8(9):71-80.

Institute of Medicine. 1992. Emerging Infections: Microbial Threats to Health in the United States. Washington, D.C.: National Academy Press.

Irving, L. G. 1982. Viruses in wastewater effluents. P. 7 in Butler, M., A. R. Medlen, and R. Morris (eds.) in Viruses and Disinfection of Water and Wastewater. United Kingdom: University of Surry.

James M. Montgomery, Inc. 1979. Water Factory 21 Virus Study. Conducted under contract to Orange County Water District, Fountain Valley, Calif. Pasadena, Calif.: James M. Montgomery, Consulting Engineers, Inc.

James M. Montgomery, Inc., 1981. Water Factory 21 Environmental Virus and Parasite Monitoring. Conducted under contract to Orange County Water District, Fountain Valley, Calif. James M. Pasadena, Calif.: Montgomery, Consulting Engineers, Inc.

James M. Montgomery, Inc. 1983. Operation, Maintenance, and Performance Evaluation of the Potomac Estuary Experimental Water Treatment Plant, Executive Summary, September 1980-September 1983. Conducted under contract to the U.S. Army Corps of Engineers, Baltimore District. Alexandria, Va.: James M. Montgomery, Consulting Engineers, Inc.

Kapikian, A. Z., M. K. Estes, and R. M. Chanock. 1996. Norwalk Group of Viruses. Pp. 783-810 in Fields, B. N., D. M. Knipe, and P. M. Howley (eds.) Fields Virology, Third Edition. Philadelphia: Lippincott-Raven Publishers.

Kaplan, J. E., R. Feldman, D. S. Campbell, C. Lookabaugh, and G. W. Gary. 1982. The frequency of a Norwalk-like pattern of illness in outbreaks of acute gastroenteritis. Am. J. Public Health 72:1329-1332.

Kappus. K. K., D. D. Juranek, and J. M. Roberts. 1992. Results of testing for intestional parasites by state: Diagnostic laboratories, United States, 1987. Morbidity and Mortality Weekly Report 40(SS-4): 25.

Klein, P. D., Gastrointestinal Physiology Working Group, D. Y. Graham, A. Gaillour, A. R. Opekun, and E. O. Smith. 1991. Water source as risk factor for *Helicobacter pylori* infection in Peruvian children. Lancet 337:1503-1506.

Koopmans, M., A. Herrewegh, and M. C. Horzinek. 1991. Diagnosis of torovirus infection. Lancet 337(8745):859.

Koopmans, M., M. Petric, R. I. Glass, and S. S. Monroe. 1993. Enzyme-linked immunosorbent assay reactivity of torovirus-like particles in fecal specimens from humans with diarrhea. Journal of Clinical Microbiology 31:2738-2744.

Korhonen, L. K., and P. J. Martikainen. 1991. Survival of *Escherichia coli* and *Campylobacter jejuni* in untreated and filtered lake water. Journal of Applied Bacteriology 71:379-382.

Korick, D. G., J. R. Mead, M. S. Madore, N. A. Sinclair, and C. R. Sterling. 1990. Effects of ozone, chlorine dioxide, chlorine and monochloramine on *Cryptosporidium parvum* oocysts viability. Applied and Environmental Microbiology 56(5):1423-1428

Kurtz, J. B., and T. W. Lee. 1987. Astroviruses: human and animal. Pp. 92-107 in Bock G., and J. Whelan (eds.) Novel diarrhoea viruses. Ciba Foundation Symposium 128. Chichester: Wiley.

Kutz, S. M., and C. P. Gerba. 1988. Comparison of virus survival in freshwater sources. Water Science and Technology 20(11/12):467-471.

Lauer, W. C. 1990. Denver's Direct Potable Water Reuse Demonstration Project. Final Report to EPA. Denver, Colo.: Denver Water Department.

Lauer, W. C. 1991. Water quality for potable reuse. Water Science and Technology 23:2171-2180.

Lauer, W. C., S. E. Rogers, A. M. La Chance, and M. K. Nealy. 1991. Process selection for potable reuse health effects studies. Journal of the American Water Works Association 83:52-63.

Lisle, J. T., and J. B. Rose. 1995. *Cryptosporidium* contamination of water in the USA and UK: A mini-review. Journal of Water SRT-Aqua 42(4):1-15.

MacKenzie, W. R., N. J. Hoxie, M. E. Proctor, M. S. Gradus, K. A. Blair, D. E. Peterson, J. J. Kazmierczak, D. G. Addiss, K. R. Fox, J. B. Rose, and J. P. Davis. 1995. A massive outbreak in Milwaukee of *Cryptosporidium* infection transmitted through the public water supply. New England Journal of Medicine 331(3):161-167.

Madore, M. S. , J. B. Rose, C. P. Gerba, M. J. Arrowood, and C. R. Sterling. 1987. Occurrence of *Cryptosporidium* oocysts in sewage effluents and selected surface waters. Journal of Parasitology 73(4):702-705.

Mascher, F., F. F. Reinthaler, D. Stunzner, and B. Lamberger. 1988. *Aeromonas* species in a municipal water supply of a central European city: Biotyping of strains and detection of toxins. Zentralblatt Bakteriologie, Mikrobiologie, und Hygiene 186:333-337.

Matsui, S. M., and H. B. Greenberg. 1996. Astroviruses. Pp. 811-824 in Fields, B. N. et al. (eds.) Fields Virology, Third Edition. Philadelphia: Lippincott-Raven Publishers.

McFeters, G. A., and S. I. Terzieva. 1991. Survival of *Escherichia coli* and *Yersinia enterocolitica* in stream water: Comparison of field and laboratory exposure. Microb. Ecol. 22:65-74.

McIntosh, K. 1996. Coronaviruses. Pp. 1095-1103 in Fields, B. N. et al. (eds.) Fields Virology. Third Edition. Philadelphia: Lippincott-Raven Publishers.

Mendall, M. A., P. M. Goggin, N. Molineaux, J. Levy, T. Toosy, D. Strachan, and T. C. Northfield. 1992. Childhood living conditions and *Helicobacter pylori* seropositivity in adult life. Lancet 339:896-897.

Millership, S. E., J. R. Stephenson, and S. Tabaqchali. 1988. Epidemiology of Aeromonas species in a hospital. Journal of Hospital Infection 11:169-175.

Mitchell, H. M., Y. Y. Li, P. J. Hu, Q. Liu, M. Chen, G. G. Du, Z. J. Wang, A. Lee, and S. L. Hazell. 1992. Epidemiology of *Helicobacter pylori* in Southern China: Identification of early childhood as the critical period for acquisition. Journal of Infectious Diseases 166:149-153.

Moore, A. C., Herwaldt, B. L., Craun, G. F., Calderon, R. L., Highsmith, A. K., Juranek, D. D. 1993. Surveillance for waterborne disease outbreaks—United States, 1991-1992. Morbidity and Mortality Weekly Report 42(SS-5):1-22.

Moyer, N. P. 1987. Clinical significance of *Aeromonas* species isolated from patients with diarrhea. Journal of Clinical Microbiology 25:2044-2048.

National Research Council. 1996. Use of Reclaimed Water and Sludge in Food Crop Production. Washington, D.C.: National Academy Press.

Nazer, H., E. Price, G. Hunt, U. Patel, and J. Walker-Smith. 1990. Isolation of *Aeromonas* spp. from canal water. Indian J. Pediatr. 57:115-118.

Oregon Health Division. 1992. A Large Outbreak of Cryptosporidiosis in Jackson County. Communicable Disease Summary, Oregon Health Division, Portland, Oregon 41(14).

Ortega, Y. R., C. R. Sterling, R. H. Gilman, V. A. Cama, and F. Diaz. 1993. Cyclospora species—A new protozoan pathogen of humans. New England Journal of Medicine 328:1308-1312.

Osterholm, M. T., K. L. MacDonald, K. E. White, J. G. Wells, J. S. Spika, M. E. Potter, J. C. Forfang, R. M. Sorenson, P. T. Milloy, and P. A. Blake. 1986. An outbreak of a newly recognized chronic diarrhea syndrome associated with raw milk consumption. JAMA 256:484-490.

Parsonnet, J., S. C. Trock, C. A. Bopp, C. J. Wood, D. G. Addiss, F. Alai, L. Gorelkin, N. Hargrett-Bean, R. A. Gunn, and R. V. Tauxe. 1989. Chronic diarrhea associated with drinking untreated water. Annals of Internal Medicine 110:985-991.

Peeters, J. E., E. A. Mazas, W. J. Masschelein, I. Villacorta Martinez de Maturana, and E. DeBacker. 1989. Effect of drinking water with ozone or chlorine dioxide on survival of *Cryptosporidium parvum* oocysts. Applied and Environmental Microbiology 55(6):1519-1522.

Pereira, H. G., A. M. Fialho, T. H. Flewett, J. M. S. Teixeira, and Z. P. Andrade. 1988. Novel viruses in human faeces. Lancet ii:103-104.

Petric, M. 1995. Caliciviruses, astroviruses, and other diarrheic viruses. Pp. 1017-1024 in Murray, P. R., and E. J. Baron (eds.) Manual of Clinical Microbiology, Sixth Edition. Washington, D.C.: ASM Press.

Rabold, J. G., C. W. Hoge, D. R. Shlim, C. Kefford, R. Rajah, and P. Echeverria. 1994. Cyclospora outbreak associated with chlorinated drinking water. Lancet 344:1360-1361.

Regli, S., J. B. Rose, C. W. Haas, and C. P. Gerba. 1991. Modeling risk for pathogens in drinking water. Journal of the American Water Works Association (November):76-84.

Robert, L. J., A. T. Campbell, and H. V. Smith. 1992. Survival of *Cryptosporidium parvum* oocysts under various environmental pressures. Applied and Environmental Microbiology 55(6):1519-1522.

Rogers, S. E., and W. C. Lauer. 1992. Denver's demonstration of potable reuse: Water quality and health effects testing. Water Science and Technology 26:1555-1564.

Rose, J. B., and C. P. Gerba. 1990. Assessing potential health risks from viruses and parasites in reclaimed water in Arizona and Florida. Water Science and Technology 23:2091-2098.

Rose, J. B., L. J. Dickson, S. R. Farrah, and R. P. Carnahan. 1996. Removal of pathogenic and indicator microorganisms by a full-scale water reclamation facility. Wat. Res. 30(11):2785-2797.

Rose J. B., M. Robbins, D. Friedman, K. Riley, S. R. Farrah, and C. L. Hamann. 1997. Evaluation of microbiological barriers at the Upper Occoquan Sewage Authority. Pp. 291-305 in 1996 Water Reuse Conference Proceedings, February 25-28, San Diego, Calif. Denver, Colo.: American Water Works Association.

Sagik, B. P., B. E. Moore, and C. A. Sorber. 1978. Infectious disease potential of land application of wastewater. Pp. 35-46 in State of Knowledge in Land Treatment of Wastewater, Vol. 1. Proceedings of an International Symposium, U.S. Army Corps of Engineers, Cold Regions Research and Engineering Laboratory, Hanover, New Hampshire.

Schubert, R. H. W. 1991. Aeromonads and their significance as potential pathogens in water. Journal of Applied Bacteriology 70:131S-135S.

Shahamat, M., C. Paszko-Kolva, H. Yamamoto, U. Mia, A. D. Pearson, and R. R. Colwell. 1989. Ecological studies of *Campylobacter pylori*. Klinische Wochenschrrift 67:62-63.

Shlim, D. R., M. T. Cohen, M. Eaton, R. Rajah, E. G. Long, and B. L. P. Ungar. 1991. An alga-like organism associated with an outbreak of prolonged diarrhea among foreigners in Nepal. Am. J. Trop. Med. Hyg. 45:383-389.

Singh, A., and G. A. McFeters. 1990. Injury of enteropathogenic bacteria. Pp. 368-379 in McFeters, G. A. (ed.) Drinking Water In Drinking Water Microbiology. New York: Springer-Verlag.

Smith, J. L., S. A. Palumbo, and I. Walls. 1993. Relationship between foodborne bacterial pathogens and the reactive arthritides. Journal of Food Safety 13:209-236.

Sykora, J. L., C. A. Sorber, W. Jakubowski, L. W. Casson, P. D. Gavaghan, M. A. Shapiro, and M. J. Schott. 1991. Distribution of *Giardia* cysts in wastewater. Water Science and Technology 24:187-192.

Terzieva, S. I., and G. A. McFeters. 1991. Survival and injury of *Escherichia coli*, *Campylobacter jejuni*, and *Yersinia enterocolitica* in stream water. Canadian Journal of Microbiology 37:785-790.

Turner, P. C., A. J. Gammie, K. Hollinrake, and G. A. Codd. 1990. Pneumonia associated with contact with cyanobacteria. British Medical Journal 300:1440-1441.

U.S. Environmental Protection Agency (EPA). 1989. Surface Water Treatment Rule, Guidance Manual for Compliance With the Filtration and Disinfection Requirements for Public Water Systems Using Surface Water Sources. Federal Register, Vol. 54, No. 124, June 29.

U.S. Environmental Protection Agency (EPA). 1996. 1977 EPA Research Grants Announcement. Report No. 600/R-96/125, , Office of Research and Development, USEPA, Washington, D.C.

van der Kooij, D. 1993. Importance and assessment of the biological stability of drinking water in the Netherlands. Pp. 165-179 in Craun, G. F. (ed.) Safety of Water Disinfection: Balancing the Chemical and Microbial Risks. Washington, D.C.: ILSI Press.

Villacorta-Martinez De Maturana, I., M. E. Ares-Mazas, D. Duran-Oreiro, and M. J. Lorenzo-Lorenzo. 1992. Efficacy of activate sludge in removing *Cryptosporidium parvum* oocysts from sewage. Applied and Environmental Microbiology 58(11):3514-3516.

Wagenknecht, L. E., J. M. Roseman, and W. H. Herman. 1991. Increased incidence of insulin-dependent diabetes mellitus following an epidemic of coxsackievirus B5. American Journal of Epidemiology 133: 1024-1031.

Weber, R., R. T. Bryan, R. L. Owen, C. M. Wilcox, L. Gorelkin, G. S. Visvesvara, and Enteric Opportunistic Infections Working Group. 1992. Improved light-microscopical detection of microsporidia spores in stool and duodenal aspirates. New England Journal of Medicine 326:161-166.

West, A. P., M. R. Millar, and D. S. Tompkins. 1990. Survival of *Helicobacter pylori* in water and saline. Journal of Clinical Pathology 43:609.

Western Consortium for Public Health. 1992. The City of San Diego Total Resource Recovery Project Health Effects Study, Final Summary Report. Oakland, Calif.: Western Consortium for Public Health.

Yanko, W. A. 1993. Analysis of 10 years of virus monitoring data from Los Angeles County treatment plants meeting California wastewater reclamation criteria. Wat. Environ. Res. 65(3): 221-226.

Yolken, R., F. Leister, J. Almeido-Hill, E. Dubovi, R. Reid, and M. Santosham. 1989. Infantile gastroenteritis associated with excretion of pestivirus antigens. Lancet I:517-519.

4

Methods for Assessing Health Risks of Reclaimed Water

Any plan to augment potable water supplies with reclaimed water must include an evaluation of the potential health risks. Yet as described in earlier chapters, such assessment is complicated by several factors, including uncertainties about the potential contaminants and contaminant combinations that may be found in reclaimed water and about the human health effects those contaminants may cause.

This chapter discusses methods and strategies for assessing the health risks of drinking reclaimed water. Previous National Research Council (NRC) reports have provided similar guidance on assessing health risks of reclaimed water (see, most recently, *Ground Water Recharge Using Waters of Impaired Quality*, published in 1994). This chapter updates and expands on information in those earlier reports. The chapter also discusses the complications of and alternative strategies for using epidemiological studies to evaluate health risks of potable water reuse.

EVALUATING MICROBIAL CONTAMINANTS

Efforts to monitor water quality for microbiological safety have historically relied on measurements of one or more groups of coliform bacteria as indicators of fecal contamination, treatment efficiency, and the integrity of the water distribution systems. Fecal coliform bacteria are indicative of fecal contamination and associated health risks; however, the measurement and control of total coliforms (rather than only fecal coliforms) during disinfection is considered to be a more stringent treat-

118

ment goal. Water quality standards have used either or both measures depending on the type of water use.

While coliform bacteria serve well as indicators of bacterial pathogens, they do not predict the inactivation or removal of enteric protozoa and viruses (Gerba and Rose, 1990; LeChevallier and Norton, 1993; LeChevallier et al., 1991a, 1991b; Rose et al., 1991). For instance, LeChevallier and Norton (1993) found that multiple linear regression models using coliforms and temperature could predict only 57 percent of the variation in *Giardia* cyst concentration, whereas no model using indicator bacteria could adequately predict *Cryptosporidium* oocyst levels.

Methods to detect viruses in water were first developed by Paul and Trask (1947) for measuring enteroviruses in untreated wastewater. Coin (1966) was one of the first to detect viruses in finished drinking water meeting the existing coliform standard. In the 1970s, improvements in collection filters and the use of antibodies led to the first isolation of rotavirus and hepatitis A virus from water. Today, standard methods for the detection of enteric viruses are based on the ability of viable enteric viruses to destroy monkey kidney cells grown *in vitro*; this cell-destroying ability is known as cytopathic effect or CPE (Benenson, 1995).

Methods for detecting microbial protozoa were first developed for *Entamoeba* in the 1940s. Starting in 1965, research focused on the detection of *Giardia*. In the 1980s, a standardized approach for *Giardia* detection was developed that used filtration for collection and antibodies labeled with a fluorescent isothiocyanate (FITC) for enhanced microscopic detection (Rose et al., 1988a, 1988b). This approach was applied to the detection of *Cryptosporidium* after its first recorded waterborne outbreak in the United States in 1985.

As more protozoan and viral waterborne outbreaks occurred in waters meeting existing water quality standards, the limitations of using indicator bacteria became apparent. In response to these health concerns, the Environmental Protection Agency (EPA) promulgated the Surface Water Treatment Rule for drinking water in 1989 (U.S. EPA, 1989a, 1989b). The rule established treatment-based performance goals of 99.99 percent reductions of viruses and 99.9 percent reduction of *Giardia*. The rule also emphasized the use of sand or multimedia filtration for the removal of *Giardia* and the use of improved disinfection methods for the control of both viruses and *Giardia*. The target reduction level was based on anticipated levels of pathogens in ambient surface waters, and the performance goals were derived from a desired annual risk of microbial disease of not greater than 1 in 10,000. An Enhanced Surface Water Treatment Rule (U.S. EPA, 1996) is under development; the enhanced rule will include an assessment of *Cryptosporidium* in source waters and its removal by treatment processes.

Arizona is the only state that has standards for the concentration of enteric viruses and *Giardia* in reclaimed water; it is also the only state in which water reuse is regulated by a laboratory certification program, although specialized studies have been undertaken in Florida and California for viruses and protozoa.

To address the lack of information nationwide on levels of viral and protozoan pathogens and the efficacy of water treatment, EPA promulgated the Information Collection Rule (ICR) in 1996. Detection processes for the enteroviruses by cell culture and for protozoa by microscopy have been standardized for this rule, and laboratories are undergoing an approval process. The data will be used in future risk analyses to establish the necessary drinking water treatment performance criteria for the protection of public health. The results will be applicable to potable reclamation projects as well. However, when considering wastewater as a source of drinking water, particular attention should be paid to current limitations and other issues involved with the methodology used to detect microbial pathogens.

Microbial Detection Methods

Microbial detection methods can be described and compared in terms of recovery (the efficiency of the method for collecting microorganisms from water samples), sensitivity (a measure of the minimum number of microorganisms that can be detected per unit volume), and specificity (the proper taxonomic identification of the microbial agent). No method is 100 percent efficient; estimates of recovery tend to range from 5 to 60 percent. A method's sensitivity is often expressed as a detection limit, such as 1/100 ml, meaning that it is able to detect one microorganism in a 100 ml sample. In untreated wastewater, concentrations of microorganisms can be high enough to be readily detected in small test volumes. However, such methods are not sufficiently sensitive for testing the highly treated reclaimed water typically produced by potable reuse projects. With highly treated reclaimed water, larger volumes of water are needed for analysis, and microorganisms may occur at concentrations too low to be detected.

Table 4-1 presents the major microbial detection techniques as they are applied to the detection and quantification of bacteria, viruses, and protozoa. Culture techniques have long been used for the detection and enumeration of viable bacteria and viruses, while microscopy techniques have a long history in the identification of bacteria and protozoa.

Table 4-2 summarizes the advantages and disadvantages of some methods for evaluating the microbiological quality of reclaimed water. The polymerase chain reaction (PCR) has only recently been applied to

TABLE 4-1 Methods and Issues for Detection of Microorganisms in Water

Microbe	Collection	Identification	Enumeration	Major Concerns
Bacteria	Low volumes: 100 to 4000 ml grab samples	Differential media and growth; biochemical testing	Colony counts; MPN	Viable but nonculturable and stressed bacteria are not accounted for. Staining and microscopy methods now available, but rarely used.
Viruses	Large volume collection: 10 to 1000 liters by filtration/adsorption	Generally, antiserum is used to identify the viruses. Most often, viruses are broadly described as either cultivable enteric viruses or enteroviruses.	PFU in cells or by an MPN method utilizing cell destruction (CPE)	Standards methods are available for enteroviruses, rotaviruses, adenoviruses, and hepatitis A virus. Only a small percentage of the viruses will cause CPE. Most studies underestimate the level of viruses.
Protozoa	Large volume: 10 to 1000 liters by filtration/entrapment	Based on antibody staining and microscopic visualization (size, shape, morphology)	Counts under the microscope	No available test for speciation or viability

NOTE: CPE = cytopathic effects; MPN = most probable number; PFU = plaque-forming units.

TABLE 4-2 Advantages and Disadvantages of Microbiological Methods for Detecting Microorganisms in Reclaimed Water

Tool	Advantages	Disadvantages	Application
Culture of bacteria and viruses	Widespread use Standardized	In some cases, lacks specificity (i.e., generic for enteroviruses; HPC)	To evaluate water quality and treatment, retrospectively For comparison to new methods and old databases
	Detects viable microorganisms	Results take days or weeks Difficult to quantitate in some cases Measures only a fraction of the types of microorganisms present; no ability to detect those that are viable but nonculturable	When viability is of major concern

| Microscopy for bacteria and protozoa | Can be used with specific stains (monoclonal antibodies, probes, fluorogenic dyes, and viability stains) Rapid, quantitative | Sample preparation can interfere with detection Less sensitive than culture as smaller volumes are examined Can be tedious | Only method readily available for protozoa Assessment of physical removal For use when stressed bacteria are an issue |
| Polymerase chain reaction | Highly specific, rapid | Measures nonviable microorganisms Can be inhibited by sample constituents Nonquantitative Less sensitive due to low volumes processed | Excellent for presence/absence assessment Only method available for key viruses of concern Can be applied with cell culture techniques for rapid and specific identification |

NOTE: HPC = heterotrophic plate count.

wastewater to identify a variety of microorganisms based on their spe-cies-specific nucleic acid sequence.

Use of Polymerase Chain Reaction Techniques

Polymerase chain reaction, or PCR, is a molecular technique used to detect a variety of microorganisms in environmental samples (Atlas et al., 1992; Bej et al., 1990; Johnson et al., 1995; Kopecka et al., 1993; Mahbubani et al., 1991; Tsai and Olson, 1991). PCR can rapidly identify a specific organism. However, before PCR can be used routinely for envi-ronmental monitoring, several issues must be addressed, including the test's sensitivity, the viability of detected pathogens, and assay interfer-ence by inhibitors in the spectrum.

The sensitivity (or the limit of detection) of PCR is constrained by the technology. In most cases, only very small volumes (100 µl or less) can be processed through the thermal cyclers (machines that control the sample temperatures during processing) used in PCR assays. Therefore, concen-tration of samples is necessary. (Alternatively, larger-capacity machines could be developed in order to increase the sample volume.)

Further, because PCR does not distinguish live from dead microor-ganisms, a cell cultivation procedure must be performed before the re-sults have relevance to health risks. This is especially true for samples taken from water that has undergone disinfection. PCR may therefore be most useful in untreated waters (source waters, recreational waters, shell-fish harvesting waters, ground waters) where viability can be assumed, or in the evaluation of processes designed to physically remove micro-biological particles (such as membrane processes).

Water quality is also an issue, since physical and chemical constitu-ents in water can mask the target nucleic acid or inhibit the enzyme reac-tion that the PCR process uses to amplify the target DNA, creating a false negative result. Recently developed antibody capture procedures appear to have great promise in addressing the problem of interference for both protozoa (Johnson et al., 1995) and viruses (Deng et al., 1994).

Finally, PCR remains only qualitative in that the results are presented as positive or negative. The development of quantitative techniques us-ing PCR would be very useful for assessments of the microbiological quality of drinking water.

Detection of Bacteria

Several well-established methods for detecting and enumerating coliform and fecal coliform bacteria indicators exist and are useful for evaluating the effectiveness of disinfection in water and wastewater treatment. Generally, any treatment process that inactivates the indicator bacteria also inactivates pathogenic enteric bacteria to similar degrees. Nevertheless, the use of indicator bacteria has limitations. For instance, recent findings suggest that very low levels (10 to 100 colony-forming units (CFU) per liter) of *Salmonella* may be related to incidences of reactive arthritis (Smith et al., 1993), suggesting that *Salmonella* should be measured directly. In addition, Singh and McFeters (1990) reported a noncultivable but viable state for *Yersinia* bacteria after disinfection. The public health significance of these noncultivable bacteria has not been fully assessed. Future regulations may dictate that greater reductions of pathogenic bacteria be achieved and documented.

Directly detecting pathogenic bacteria has traditionally been a tedious process, requiring many biochemical tests to identify the genus and/or species. Most testing has been done with presence/absence tests or the most probable number (MPN) approach. A standard MPN procedure has been developed for detecting *Salmonella* in sludge (U.S. EPA, 1992); however, the sensitivity of the test has been questioned (National Research Council, 1996). No standard procedure exists for *Salmonella* testing in reclaimed water. As described above, PCR is a rapid detection method that can be combined with more traditional cell culture techniques to assess viability of bacterial pathogens. This approach has been successfully applied in the food industry (Fung, 1994) and may hold promise for use in reclaimed water as well.

Detection of Protozoa

Detecting low concentrations of protozoan cysts or oocysts in highly treated reclaimed water requires passing large volumes of water through a filter with an appropriate pore size (typically 1.0 µm). Unfortunately, this method also concentrates unwanted constituents in the sampled water (e.g., particulates, precipitated minerals), and these constituents must then be separated in the subsequent analysis using a concentration and clarification procedure (LeChevallier and Trok, 1990; Rose et al., 1989). The semipurified sample, consisting of the larger organisms with some of the unwanted constituents removed, is then mixed with fluorescent-tagged monoclonal antibodies specific to the cyst and oocyst wall using an indirect fluorescent antibody (IFA) procedure. The sample can then be examined and the protozoa identified by one or more techniques, such as epifluorescence or differential interference contrast (DIC) microscopy (LeChevallier et al., 1991a).

The efficiency of cyst/oocyst recovery for *Cryptosporidium* and *Giardia* is quite variable, and losses occur at each of the various steps of the detection process; thus many of the organisms contained in the original water sample will be lost. Reported efficiencies are: 88-99 percent for sample collection; 16-78 percent for filter elution; 66-77 percent for concentration and clarification; and 9-59 percent for microscopic detection using the IFA method (LeChevallier and Trok, 1990; LeChevallier et al., 1991a; Ongerth and Stibbs, 1987; Rose et al., 1988b, 1989, 1991b). Given these variations in recovery, current methods for detecting protozoa tend to underestimate the true concentrations in environmental samples.

Indirect fluorescent antibody techniques using antibodies labeled with fluorescent isothiocyanate have greatly enhanced the ability to detect *Cryptosporidium* and *Giardia* in environmental samples. However, most antibody techniques provide no species identification (e.g., bird versus mammalian isolates), nor do they determine whether cysts and oocysts are viable. LeChevallier et al. (1991a, 1991b) reported that 10 to 30 percent of the organisms detected by IFA were not empty and therefore viable. Other inaccuracies may occur due to background fluorescence from naturally fluorescing organisms and from nonspecific binding of the antibody. Such problems may produce either false positives or false negatives. Although false positives can create inconvenience or necessitate further testing, Clancy et al. (1994) noted that false negatives (failing to detect protozoa that are present) pose more serious problems.

A number of IFA systems have been developed for *Cryptosporidium* and *Giardia* (Garcia et al., 1987; Rose et al., 1989; Stibbs et al., 1988). New methods under development should allow a better assessment of enteric protozoa. These methods include the use of cell cultures to detect viable and infectious oocysts, the use of immunomagnetic separation (IMS) techniques to enhance the recovery of cysts/oocysts (Jakubowski et al., 1996; Linquist, 1997; Slifko et al., 1997), and the use of internal stains for improved identification and detection.

Detection of Viruses

Viruses may be concentrated from water using either electropositive or electronegative filters. Either can concentrate viruses from large volumes of water, but clogging, particularly in positively charged filters, may occur if the water is high in suspended solids or turbidity (Rose et al., 1989). The adsorbed viruses are eluted and concentrated to a smaller volume using an organic flocculation procedure. This final concentrate may be inoculated into cell cultures to detect cytopathic effects, or it may be tested by PCR or other types of tests.

Detection methods have primarily been developed for the enterovirus group, which consists of poliovirus, echoviruses, and coxsackieviruses. Microbiological studies in the San Diego potable reuse project demonstrated the feasibility of recovering 1 virus in 1000 liters of water.

When large volumes of water are sampled, one of the unresolved technical issues is whether a given filter will adsorb an equivalent amount of viruses per unit volume of water as more water is passed through the filter or whether the filter's performance will decrease as the quantity of water sampled increases. Further, some research has suggested that not all enteroviruses are adsorbed by filters at the same rate (Powelson and Gerba, 1995).

Cell culture techniques using CPE (cytopathic effect, or cell destruction) on monkey kidney cell lines provide the principal method for virus detection. In contrast to processing procedures for detecting protozoa, the debris collected along with viruses is more easily separated before cell culture. The types of viruses are further identified by the use of specific antibodies to neutralize the virus's cytopathic effect. However, these cell culture methods may detect only a small percentage of the viruses found in polluted waters because not all viruses will cause CPE. When Payment and Trudel (1987) used labeled antibodies to determine the numbers of infectious foci in the cell culture, they found levels of viruses up to 100 times greater than what they detected by CPE.

While enteroviruses are the most heavily studied waterborne viruses, the limited data on virus identification indicate that retroviruses typically occur at higher levels in wastewater (Gerba and Rose, 1990). Hurst et al. (1988) found that adenoviruses could be detected at levels 94 times higher than enteroviruses when using molecular techniques (gene probes) to verify the results of cell cultures.

In sum, although methods exist for detecting some viruses, we lack sufficient data to determine the concentrations and diversity of the many viruses of concern in wastewater, reclaimed water, and ambient waters (Hurst et al., 1989). These include emerging viruses of concern such as Norwalk viruses.

Indicator Techniques

Several types of microorganisms have been suggested as alternatives to coliform bacteria as indicators of water quality, fecal pollution, and public health risks. These include the bacteria *Enterococcus* and *Clostridium perfringens* and the F-specific coliphage bacterial virus.

The *Enterococcus* bacteria comprise a subgroup of the fecal *Streptococci* bacteria and include *S. faecium*, *S. faecalis*, *S. durans*, and related biotypes (Clausen et al., 1977). *Enterococci* are generally more resistant to water treatment than bacterial pathogens or fecal coliforms, and membrane filtration procedures are available for sampling them (Cohen and Shuval, 1973; Davies-Colley et al., 1994; Sinton et al., 1994). Some epidemiological investigations found that levels of *Enterococcus* correlated with an increased incidence of gastrointestinal symptoms associated with exposure to polluted recreational waters (Cabelli et al., 1979; Dufour, 1984). The EPA has suggested that *Enterococcus* may be the best microbial indicator for ambient waters (U.S. EPA, 1986). A few states have adopted the use of *Enterococcus*, but most continue to use total and/or fecal coliforms for evaluating recreational water quality and the wastewater effluents that may impact these waters.

Clostridium perfringens is a pathogenic bacterium found in human and animal feces. Because it forms resistant spores, *Clostridium* has been recommended as a conservative indicator of water quality and as a valuable supplement to other water quality tests, particularly in situations where the detection of viruses or fecal contamination is desirable (Fujioka and Shizumura, 1985; Payment and Franco, 1993). This microorganism is consistently present in municipal wastewater at concentrations of 10^3 to 10^4 CFU/100 ml. Fairly rapid and simple membrane filtration methods are available for the enumeration of *C. perfringens*. Researchers have found a significant correlation between levels of *C. perfringens* and enteric viruses and protozoa in evaluations of the treatment efficiency of filtration and disinfection (Fujioka and Shizumura, 1985; Payment and Franco, 1993). These authors suggested that the removal of this indicator by a factor 7 to 8 $\log_{10}$ essentially ensures the removal of enteric protozoa and viruses at current limits of detection. However, no jurisdictions in the United States are known to have adopted *Clostridium* as a regulatory standard (Cabelli, 1997).

The coliphages are viruses that infect *Escherichia coli*, and therefore the presence of coliphages in water indicates the presence of their host *E. coli*, which is excreted by animals and humans. Coliphages may serve as better indicators for human enteric viruses than bacterial indicators do because coliphages more closely resemble human enteric viruses in size, shape, and resistance to treatment processes. In a comparison of untreated and treated wastewater, river water, treated river water, and treated lake water, Havelaar et al. (1993) found significant correlations between levels of coliphage and levels of enteric viruses. However, this correlation was not evident for the untreated and treated wastewater samples, which suggests that other unknown factors may complicate the use of this indicator when evaluating recent wastewater inputs into a water body.

Payment and Franco (1993) examined the removal of coliphages and enteroviruses in drinking water treatment. The total removal and/or inactivation of enteroviruses by the complete drinking water process was estimated at 7 $\log_{10}$, based on coliphage removal. Coliphages have also been used as biological tracers in the environment to evaluate the movement of septic tank effluent (Paul et al., 1995). An advantage of using coliphages to evaluate treatment processes is that large concentrations can be seeded into pilot or full-scale systems, and low numbers can be monitored after dilution and reduction by treatment to indicate process removal (Rose et al., 1996). Such seeded studies may provide useful data more rapidly than monitoring of indigenous microorganisms can.

Microbial Risk Assessment

Human health risk assessment is a scientific process that attempts to identify and quantify the health risks from exposure to environmental hazards. The use of risk assessment in drinking water began in 1974 with the congressional mandates of the Safe Drinking Water Act. A series of studies and reports from the National Research Council evaluated health risks from contaminants in drinking water (National Research Council, 1977-1989). The EPA used the results of these studies and risk assessment to set maximum contaminant levels (MCLs) for a number of inorganic and organic chemicals. The Surface Water Treatment Rule (U.S. EPA, 1987) used a risk assessment approach, based on the presumed levels of pathogens in surface waters, to mandate certain treatment processes designed to reduce the annual risk of disease occurrence from viruses and protozoa to less than one disease occurrence per 10,000 people served (Regli et al., 1991).

Microbial risk assessments have been performed for nonpotable reclaimed water systems based on hypothetical exposure scenarios (Asano et al., 1992) and on monitoring data (Rose et al., 1996), but no approach has been formally developed for reclamation systems used for potable water supplies.

Recently, a conceptual framework for evaluating the risks associated with exposure to microbial contaminants has been developed (ILSI, 1996). This approach modifies a four-step paradigm developed by the National Research Council for chemical contaminants (see Box 4-1) into one better suited to microbial contaminants. The first phase of the process involves characterizing the human health effects and determining the dose-response relationship. This information comes from clinical, epidemiological, and public health surveillance data. The second phase involves estimating exposure and using the dose-response models to quantify the health risk. This two-phase approach is described below.

Human Health Effects and Dose-Response Modeling

Microbial risk assessment requires adequate knowledge about the relationship between exposure to microorganisms and consequent health effects. Clinical and epidemiological studies provide the information nec-

---

BOX 4-1
Steps in Risk Assessment

In 1983 the National Research Council defined four steps in the risk assessment process:

1. hazard identification, involving definition of the human health effects associated with any particular hazard;
2. dose-response assessment, involving characterization of the relationship between the dose administered and the incidence of the health effect;
3. exposure assessment, involving determination of the size and nature of the population exposed and the route, amount, and duration of the exposure; and
4. risk characterization, or integration of the three steps in order to estimate the magnitude of the public health problem.

These guidelines have generally been followed in evaluating the health risks of specific chemicals in the environment and, with some modification, have been used for microbial contaminants.

---

essary to determine the dosage or exposure level necessary to cause infection or disease. Infection is the colonization of microorganisms in the body and may or may not cause symptoms. Infectious dose is often expressed as the dose required to cause an infection in 50 percent of the population who are exposed (ID_{50}).

Infectious dose has been evaluated in human volunteers for many enteric microorganisms in order to develop dose-response curves. Haas (1983) first developed dose-response models using data from human volunteers who ingested various levels of viruses, protozoan cysts or oocysts, or bacteria. The individuals were then monitored for infection and, in some cases, disease. The ratio of those infected to those exposed at each dose formed the basis for the dose-response curve.

Some recently developed models provide a good fit to human dose-response data sets. Haas et al. (1993) found that a beta-Poisson model best described the probability of virus infection; Rose et al. (1991) and Haas et al. (1996) also developed exponential models to evaluate the risks of *Giardia* and *Cryptosporidum* infections. Table 4-3 summarizes the dose-response models for six different microorganisms from various studies. Estimates of risk may be obtained by substituting a given value for exposure in the table, represented as *N* or number of microorganisms ingested.

TABLE 4-3 Probability of Infection Models and Best Fit Dose-Response Parameters for Various Human Feeding Studies

Organism	Best Model[a]	Model Parameters[b]
Rotavirus	Beta-Poisson	$\alpha = 0.26$ $\beta = 0.42$
Giardia	Exponential	$r = 0.0198$
Salmonella	Exponential	$r = 0.00752$
E. coli	Beta-Poisson	$\alpha = 0.1705$ $\beta = 1.61 \times 10^6$
Shigella	Beta-Poisson	$\alpha = 0.248$ $\beta = 3.45$
Cryptosporidium	Exponential	$r = 0.00467$

[a]Models

$P_i \quad = \quad 1 - (1 + N/\beta)^{-\alpha}$ (beta-Poisson)

$P_i \quad = \quad 1 - \exp(-rN)$ (exponential)

[b]Model parameters:

P_i = probability of infection (ability of the organism to establish and reproduce in the intestine)

N = exposure, expressed as numbers of microorganisms ingested (CFU of bacteria, cysts, or oocysts of Giardia or Cryptosporidium, or PFU (plaque-forming units) of viruses)

α, β, r = constants for specific organisms that define the dose-response model

SOURCE: Haas et al., 1996; Rose et al., 1991a.

Exposure and Risk Characterization

Using previously defined models for infection and data from human health effects, one can estimate the health risk a pathogen poses at various exposures. Table 4-4 compares four microbial contaminants for risks of infection, severity of symptoms, mortality in normal populations, and mortality in vulnerable populations. The values for exposure in drinking water were estimated based on concentrations reported in surface waters and using treatment reductions of 99.9 percent for *Giardia*, 99.99 percent for rotavirus, and 99.9999 percent for *Salmonella*. With these assumptions, the risk of infection ranged from a low of 2.7×10^{-4} to a high of 4.9×10^{-3}, depending on the microorganism. While the dose models shown earlier (Table 4-3) indicate that the infectivities of rotavirus, *Giardia*, and *Crypto-*

TABLE 4-4 Comparative Risks of Disease and Mortality From Infections Acquired via Drinking Water

Etiologic Agent	Daily Exposure (number of organisms)[a]	Daily Risk of Health Outcome			
		Infection[b]	Severity[c]	Mortality[d]	Vulnerable Mortality[e]
Rotavirus	0.008	4.9×10^{-3}	4.4×10^{-5}	4.0×10^{-7}	4.9×10^{-5}
Salmonella	0.116	2.7×10^{-4}	1.1×10^{-5}	2.7×10^{-7}	1.0×10^{-5}
Giardia	0.132	2.6×10^{-3}	1.8×10^{-5}	2.6×10^{-9}	f
Cryptosporidium	0.96	4.5×10^{-3}	4.1×10^{-5}	4.1×10^{-9}	$2.1 \times 10^{-5}g$

[a] These are considered as high exposure estimates based on treatment reductions of 99.9% for Giardia, 99.99% for rotavirus, and 99.9999% for Salmonella, and assuming consumption of 2 liters of water per day.

[b] Daily risk of infection is based on the probability of infection modeled for the specific microorganism and the corresponding daily exposure given in the table.

[c] Risk of severity is calculated as the number of individuals hospitalized, divided by the total number of cases for the particular microorganism that have been documented during waterborne outbreaks, and then multiplied by the probability of infection (taken from Rose et al., 1996).

[d] General population.

[e] Nursing homes.

[f] No documented increase in mortality in the nursing home population.

[g] Risk in immunocompromised.

sporidium vary, their overall risks of infection are similar because of the differences in their potential exposure. The bacteria, represented by *Salmonella*, represent the lowest risk of infection. When evaluating mortality rather than just infection, the bacteria and viruses become more significant (by a hundredfold) than protozoa.

Risks from microbial contamination depend not only on the dose of microorganisms ingested but also on the host's immune status. Sensitive populations, including the elderly, infants, and people with compromised immune systems, stand at greater risk of severe outcomes (Gerba et al., 1996). For example, case-fatality ratios go from 0.001 percent for *Cryptosporidium* in the general population to 50 to 60 percent in the immunocompromised population (Rose, 1997). The immunocompromised population's risk of mortality is 100 times greater for viruses and bacteria and 10,000 times greater for *Cryptosporidium*.

An individual who contracts an infection from exposure to a waterborne agent may in turn serve as a source of infection for other individuals regardless of whether symptoms are apparent. These secondary infections should be considered when assessing the risk of a particular exposure.

Such secondary cases may arise by a variety of mechanisms. Among close family members, household secondary cases can arise by direct or indirect (such as from surface or food contamination) contact; this is particularly a concern when the infected individual is a child (Griffin and Tauxe, 1991; MacKenzie et al., 1995). Table 4-5 summarizes secondary transmission statistics obtained from a variety of waterborne and non-waterborne outbreaks. It is unclear whether these secondary statistics would be applicable in nonoutbreak situations.

The spread of infectious disease in a population may be modeled using standard population-state models from mathematical epidemiology (Bailey, 1975, 1986). When applied to the transmission of waterborne infectious disease, such models contain many parameters, including some for which only poor or no estimates are available. However, a preliminary illustration of the methodology as applied to reclaimed potable water does exist (Eisenberg et al., 1996). These models account for typical numbers of infected, ill, recovering, and immunocompromised subpopulations in computing the passage of an infectious disease through a community.

Actual levels of exposure to microorganisms through drinking water have been difficult to estimate, because the data on reductions of many pathogens are limited and extrapolations are often needed to determine tap water concentrations. A limited amount of data on virus and protozoa removals during water treatment is available (LeChevallier et al., 1991a; Payment and Franco, 1993; Payment et al., 1985; Stetler et al., 1984).

TABLE 4-5 Summary of Secondary Case Data in Waterborne Disease Outbreaks

Organism	Secondary Attack Ratio[a]	Secondary Prevalence in Households[b]	Source of Outbreak
Cryptosporidum parvum	0.33	0.33	Contaminated apple cider (Millard et al., 1994)
C. parvum	N/A[c]	0.042	Drinking water in Milwaukee (MacKenzie et al., 1995)
Shigella	0.28	0.26	Child day care center (Pickering et al., 1981)
Rotavirus	0.42	0.15	Child day care center (Pickering et al., 1981)
Giardia lamblia	1.33	0.11[d]	Child day care center (Pickering et al., 1981)
Viral gastroenteritis	0.22	0.11[d]	Drinking water in Colorado (Morens et al., 1979)
Norwalk virus	0.5-1.0	0.19	Swimming (Baron et al., 1982)
Escherichia coli O157:H7	N/A	0.18[d]	Child day care center (Spika et al., 1986)

[a]Ratio of secondary cases to primary cases.
[b]Proportion of households with one or more primary cases who have one or more secondary cases.
[c]N/A = information not available.
[d]Proportion of persons in contact with one or more primary cases who have a secondary case.

Nevertheless, the lack of monitoring data for evaluating exposure remains the greatest barrier to adequately assessing risks posed by microbial pathogens.

A final difficulty in assessing risks from microbial contamination of reclaimed water is in evaluating the environmental fate of pathogens in indirect potable reuse projects. As explained in Chapter 2, pathogens die slowly under certain environmental conditions, with the rate of die-off dependent primarily on temperature. However, monitoring the fate of pathogens in the environment is extremely difficult due to dilution and a lack of knowledge of the hydrology of the receiving water.

These complicating factors introduce great uncertainty in the assessment of potential risks due to microbial contamination of reclaimed water. System reliability is therefore crucial. As discussed in Chapter 6, the reliability of treatment processes in reducing risks will need to be critically assessed, and multiple treatment barriers will be needed in case one of the processes fails. Further, public health officials involved with surveillance programs (also discussed in Chapter 6) in areas using reclaimed water will need to be aware of the symptoms associated with infections arising from waterborne pathogens, including emerging pathogens, and may need to broaden the scope of public health monitoring.

EVALUATING THE RISK OF CHEMICAL CONTAMINANTS

Evaluating the risk posed by chemical contaminants poses challenges in any context, and some unique challenges in reclaimed water.

Safety Evaluations of Reclaimed Water

Calculations of health risk from chemicals in drinking water are largely based on extrapolations of the results of toxicological experiments on animals and estimates of human exposure to the chemical. Based on such risk information, national standards have been established for a limited but growing number of chemical contaminants in drinking water. This approach has been considered adequate for drinking water derived from relatively pristine sources or from sources that have been used for a long time without evidence of harm. In this sense, the regulation of drinking water resembles the approach used for establishing the safety of commercially produced foods, in that it assumes the use of an approved (and relatively safe) source of raw materials, then concentrates on monitoring for common contaminants in these materials or for problems in their processing (e.g., a lack of good manufacturing practices).

However, current drinking water standards are not intended to ensure the safety of reclaimed water because the wastewater may introduce new, unknown, or unquantified sources of contamination (as explained in Chapter 2). For instance, we have poor toxicological data on the wide variety of organic compounds in wastewater. Thus, even if a water reuse system could identify all the organic material in its processed wastewater (which, as explained in Chapter 2, is impossible), there would be no basis for assigning risks to most of the identified compounds. Because of this and because many compounds are unidentifiable, toxicological testing of reclaimed water may be the only way to ensure the water's safety. However, applying toxicological methods to reclaimed water is difficult because of the water's variable chemical composition and the difficulty of obtaining representative samples. The challenge becomes how to test high doses of essentially unknown contaminants so that standards with some margin of safety can be established.

The chemical hazards in reclaimed water derived from municipal wastewater may vary depending upon a number of factors, including (1) the composition of chemicals intentionally or unintentionally discharged to the system from domestic and industrial sources, (2) variability in the removal of these chemicals by treatment, (3) the introduction of chemicals during the treatment process, and (4) the creation of entirely new chemicals as a result of chemical reactions during the treatment. If all wastewaters had the same composition, one could extrapolate the apparent safe consumption of a drinking water augmented with treated wastewater from one community to another. However, the complexity and potential variability in the chemical composition of wastewaters from different localities make such extrapolation quite difficult.

Recent research on the toxicity of different chemical mixtures has shown that the main health hazards of a chemical mixture are frequently posed by very low concentrations of highly potent chemicals (National Institutes of Health, 1993). The potencies of chemicals as toxicants or as carcinogens easily span 6 to 9 orders of magnitude. Therefore, toxic effects of chemicals in a wastewater are not necessarily attributable to the chemicals that occur in the highest concentrations.

Another factor in evaluating uncharacterized mixtures from treated wastewater is the presence of large amounts of nontoxic chemicals that complicate the testing of mixtures. For example, most drinking water contains inorganic salts that pose absolutely no health concern at the concentrations at which they occur. Yet these salts must be separated out before the water's organic chemicals can be evaluated, because concentrating the salts would dehydrate and kill experimental animals in the same way that sea water would.

The earlier National Research Council review *Quality Criteria for Water Reuse* (NRC, 1982) made some recommendations related to the type of testing that should be applied to establishing the safety of reclaimed wastewater. Much of the remainder of this section assesses the adequacy of that approach. Of particular importance is the evolution of how the results of specific tests are interpreted, how they fit in an overall scheme of health-effect testing procedures, and how decision rules (sometimes unstated) are applied to the results of tests.

Checking the reverse osmosis unit at San Diego's Aqua III water reclamation facility. Photo by Joe Klein.

Approaches to Toxicological Testing

The most common way to evaluate chemical hazards to human health is to test compounds in live animals (usually mammals), a process known as *in vivo* testing. For special purposes, testing of lower life forms, such as bacteria in cell cultures, is used; this is known as *in vitro* testing. Food and drug industries have developed fairly standardized testing strategies of both types to detect adverse health effects of a given chemical with some level of certainty. First, screening tests identify which organs and/or physiological functions are most sensitive to the effects of the chemical. Then more specialized investigations assess the precise effects on physiological functions and seek to establish dose-response relationships. Conducted rigorously, such testing yields information that can be accurately extrapolated to humans.

In general, *in vivo* testing is the most effective screening method because mammals possess all the physiological systems that can be affected by a chemical in humans, and gross effects can be readily identified. *In vitro* investigations usually seek to evaluate the mechanism of a chemical's impact rather than its actual effects or, more recently, to provide direct sensitivity comparisons of human cells relative to those of the animals in which prior screening work has been done. Usually such information must be related back to the whole animal or human to determine how much active metabolite of a chemical reaches the receptor site at a given dose. Ideally, dose-response relationships are then developed to compare the internal concentration of the chemical resulting from external exposure of humans and experimental animals.

In vivo testing provides the best screening data but requires considerable investment. For example, experiments designed to detect the carcinogenic properties of a chemical generally involve treatments over a significant fraction of a test animal's life span. Reproductive or developmental effects require specific assessments of reproductive competence of sexually mature animals or observations of development in pregnant animals, which in turn requires observations over more than one generation.

Because of the expense involved in live animal testing, many *in vitro* tests have been developed for screening purposes with the hope that they would be predictive of carcinogenic and reproductive effects. These *in vitro* screening tests are not intended as a substitute for *in vivo* testing, but merely to identify chemicals needing further testing.

In the 1970s, a series of inexpensive *in vitro* tests was introduced into common use in safety testing. The most widely used was the *Salmonella/* microsome assay, commonly known as the Ames test. Its purpose was to test for mutagenic activity, because mutation plays an important role in the development of cancer. This test was subsequently applied to a wide variety of environmental problems, including drinking water and reclaimed water, in the hopes that it would provide a cost-effective method for evaluating carcinogenic hazards in the environment. The apparent success of the test spawned an interest in developing *in vitro* techniques to detect other toxicological end points.

As illustrated in Table 4-6, a safety testing strategy emerged that in-

TABLE 4-6 Progression in Safety Testing

Screening/ Hazard Identification	Dose-Response Determination	Risk Assessment

Short-term *in vitro*
  Point mutation
  Clastogenesis
  Reproductive effects
  Developmental effects
  Neurotoxicity
  Immunotoxicity
  Target organ testing
 	
Short-term *in vivo*
  Mutagenesis
  Clastogenesis
  Carcinogenesis
  Target organ identification
  Developmental effects
  Reproductive effects
  Neurotoxicity
  Immunotoxicity
  Endocrine disruption
  Metabolism effects
	Controlled human experimentation (*in vivo* can only be done in special circumstances, e.g., drug development)
	
	Animal studies, including two species tests for specific end points and studies of metabolism and pharmacokinetics
		Epidemiological studies
		
		Determination of human exposure
		
		Application of conventional risk assessment techniques to *in vivo* data

corporates short-term *in vitro* and *in vivo* testing for screening purposes. Broadly speaking, the purpose of screening is to make sure that no significant health effect is overlooked. These testing schemes seek to establish a finite, cost-effective set of tests that can quickly and inexpensively identify the chemicals that pose enough possible risk to warrant more extensive testing. Depending the nature of the chemical, the extent to which prior toxicological data are available, and the chemical's intended use, a chemical may undergo testing by some or all of the screening tests listed in Table 4-6. For example, an organophosphorus pesticide (which works by interfering with neurological functions) would be tested both *in vivo* and *in vitro* for its ability to produce delayed neurotoxicity. If the screening tests suggest the compound has a toxic potential, then careful dose-response studies would be conducted. Further testing is especially important if a chemical seems capable of producing specific types of toxicity (e.g., neurotoxicity, reproductive toxicities, teratogenesis) at dose levels that are not lethal to the animal being tested. Generally the tests are run on at least two species to ensure that no large interspecies differences in responses are distorting the test results.

Conventionally, risk assessments are made from these data, with appropriate uncertainty factors included to compensate for lack of direct data in humans. It is occasionally possible to make quantitative comparisons of internal dosimetry between experimental animals and humans if appropriate *in vitro*-to-*in vivo* and *in vitro*-to-*in vitro* comparisons of metabolism and pharmacokinetics of the chemical exist. These data, in turn, can be used to improve the accuracy of a conventional risk assessment procedure, which typically incorporates large uncertainty factors.

This general strategy has a history of use in the testing of food additives and new therapeutic drugs. The EPA has used the strategy under the Federal Insecticide, Fungicide, and Rodenticide Act and the Toxic Substances Control Act, and the Food and Drug Administration (FDA) uses the process as well. The general practice has been to test such products to the point where there is clear evidence of some toxic effect, then compare the toxic doses to the levels normally used in food or drugs to ensure an appropriate margin of safety.

The Role of Toxicological Testing in Water Reclamation Projects

The 1982 National Research Council evaluation of health considerations for the potable use of reclaimed water found that it was a practical impossibility to identify and test the toxicity of all of the individual compounds present in reused water, and that it was ultimately necessary to test the toxicity of mixtures of chemicals instead. Because the available toxicological tests were insensitive to the low concentrations present in reclaimed water, it was recommended that the mixtures be concentrated so that chemicals would be present in concentrations high enough to detect effects. Recognizing the complex and essentially unverifiable nature of the mixtures involved, the 1982 report recognized that the following factors complicate the experimental results of any toxicological testing:

- The changing consistency of samples can be dealt with only by frequent resampling and testing.
- Additive, synergistic, or antagonistic effects of the mixture components may vary with individual samples and change with time.
- The concentration procedures could influence the chemical or physical composition of samples.
- The chemical and physical stability of concentrates is essentially an unknown.
- The mechanics of sample preparation for administration to animals may influence results.

TABLE 4-7 Toxicological Tests Recommended by the 1982 NRC
Report *Quality Criteria for Water Reuse*

Phase I	Phase II	Phase III
In Vitro		
Mutagenicity		
In vitro transformation		
In Vivo		
Acute toxicity	Subchronic 90-day study	Chronic lifetime feeding
Teratogenicity	in at least one rodent	study in one species of
Short-term repeated	species, preferably in two	rodent
dose studies—14 days	species	
(includes cytogenetics	Reproductive toxicity	
assay)		

SOURCE: NRC, 1982.

Notwithstanding these complications, and based on the general testing logic outlined in Table 4-6, the 1982 NRC panel recommended that a
final comparison should be made between reused and conventional water based on the outcomes of a series of tiered tests designed to give
information on the relative toxicities of the concentrates from the two
water supplies (Table 4-7). Phase I of this recommended protocol includes *in vitro* assessments of mutagenic and carcinogenic potential by
means of microbial and mammalian cell mutation and *in vivo* evaluations
of acute and short-term subchronic toxicity, teratogenicity, and clastogenicity. Phase II includes a longer term (90-day) subchronic study and a
test for reproductive toxicity on live animals. Phase III is a chronic lifetime animal feeding study.

Several studies of reclaimed water have included toxicological testing of organic concentrates derived from water. Unfortunately, in most
of these studies the selected screening-level tests that correspond to
NRC's Phase I were not always followed up with the more detailed Phase
II and III studies necessary to confirm and elaborate on the results of the
screening tests. As shown in Chapter 5, much of the testing was limited
to *in vitro* tests designed to detect specific end points (e.g., mutagenicity
and cytotoxicity). The value of *in vitro* testing is unclear when used alone
to examine low-level risks in the environment. Much of the problem
arises from inadequate follow-through on the decision rules of safety
evaluations. These issues are explored in more detail in the following
discussion.

Limitations of *In Vitro* Testing

There are several reasons why *in vitro* testing alone does not reliably measure health risk. As used in product safety testing, the original vision for mutagenicity tests was that they would have a low false negative rate for carcinogenic compounds and that any false positive data would be revealed by further testing. In practical terms, if a test is positive, a decision can be made either to drop the product altogether or to test further with systems that better assess the chemical's carcinogenicity.

However, several studies have shown the Ames test to be a poor predictor of carcinogenicity in general (Ashby and Tennant, 1988; Douglas et al., 1988; Parodi et al., 1983). A particular concern is a higher than anticipated false negative rate in these tests that has not been improved significantly by combining different types of tests for mutagenic effects.

To be useful (i.e., to provide information that cannot be gained in another way), *in vitro* test systems must focus on a single end point (e.g., carcinogenicity, mutagenicity, teratogenicity, reproductive toxicity, liver toxicity, neurotoxicity, kidney toxicity) or, in some cases, on one of several modes of action that can produce an end point (e.g., on mutagenesis, which can produce cancer). No single *in vitro* testing system addresses a substantial fraction of the many potential health effects that chemicals can have on animals or humans, and there seems little hope that any combination of *in vitro* tests will do so or be able to substitute for experiments on live animals in the foreseeable future. Negative results in a single mutagenesis test, for instance, provide no confidence that the chemical or product causes no nerve damage, liver damage, reproductive problems, and so on.

In vitro test systems do have considerable value as tools to resolve questions of extrapolation, for they allow very directed studies of mechanisms of action. However, such studies assume that some specific type of pathology has already been identified in an intact animal or human. Therefore, the toxicological use of *in vitro* systems is moving away from screening toward risk assessment of identified contaminants. The results of such risk assessment studies are used to select the extrapolation model rather than to provide data that would be used in a quantitative fashion.

These limitations raise the question of whether screening tests for specific end points should involve *in vitro* tests at all. Only in intact animals can all forms of toxicity be recognized in one test system. This becomes even more important when the material being tested is a complex mixture such as reclaimed wastewater. Using a large battery of tests to identify each and every potential toxic effect of a complex chemical composition that varies in time and place would be very expensive and time-consuming and produce results of questionable utility. Moreover, testing must eventually return to an *in vivo* system before *in vitro* data can be related in any plausible way to actual risk.

Extrapolation of Test Results to Human Risk

Toxicological investigations of food additives or drugs make extrapolations from experimental systems to the human system and from the high experimental doses to the lower doses that are usually encountered by humans. Extrapolation from the high doses used in toxicological experiments to the low doses that might be present in reclaimed water creates uncertainty because effects that occur at high doses may not occur at low doses. In the case of presumed stochastic responses, particularly those that arise from chemically induced mutation, an assumption is sometimes made that low doses create the same responses as high doses, but with less frequency or severity—that is, that the response curve is linear. But this assumption ignores the role of nonmutagenic processes that occur only at high doses, such as those used in studies in experimental animals. These problems can be addressed by research on individual chemicals. However, the data that would allow reasonable quantitative conclusions to be made are still quite limited and do not clearly apply to the highly complex mixtures of chemicals found in wastewaters.

For these reasons and others, a recent workshop on molecular and cellular approaches to risk assessment concluded that the current state of knowledge does not allow meaningful risk assessments to be made purely on *in vitro* data (Sutter, 1995). Workshop participants recommended parallel *in vitro/in vivo* experiments in animals to validate an *in vitro*-to-*in vivo* extrapolation. *In vitro* measurements should then be made for critical end points in the appropriate human cells before the results can be extrapolated to humans (see Figure 4-1).

When considering potable reuse of treated wastewater, extrapolations to other locations and forward through time may also be necessary. While the few fairly extensive toxicological studies of reclaimed water that have been conducted (see Chapter 5) have shown no toxicological effects produced by reclaimed water, the extrapolation of these results to other reuse systems is greatly complicated by possible differences between municipal wastewaters from different systems, potential changes in wastewater composition over time, and the types of monitoring that would be needed to ensure continued toxicological safety of the reclaimed water.

Further complicating these extrapolations are the many chemicals present at low dose levels in drinking water from either natural or wastewater-influenced sources, which may interact with one another in an additive, synergistic, or antagonistic manner to produce biological effects

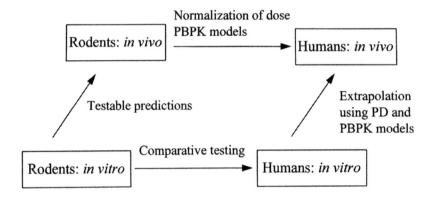

FIGURE 4-1 Parallelogram approach to risk assessment from whole-animal and *in vitro* testing procedures. The essential features are to confirm the critical toxicological response at the *in vitro* level and to adjust predictions of risk for humans based upon confirmation of *in vivo* observations in the appropriate target cell in intact animals. Development of the dose-response curve for humans would then involve modeling of target site dose with physiologically based pharmacokinetic (PBPK) models and response by pharmacodynamic (PD) models.

(NRC, 1989). Further, individuals within an exposed population vary in their susceptibility to chemical-induced injury, depending on factors such as age, nutritional status, gender, or genetic predisposition. Immunocompromised individuals, such as those undergoing organ transplant treatments or cancer chemotherapy or those infected with HIV, may suffer further immunosuppressive effects by exposure to some chemicals in drinking water at even very low dose levels. Given the millions of people using public drinking water supplies, it is reasonable to expect that such scenarios currently exist, and they should be factored into risk assessment strategies for treatment of public water supplies from any origin.

There is often limited information regarding how closely a sample used for testing represents the original constituents (see Chapter 2). Further, there has been little follow-up on the relationship between *in vitro* bioassays and live animal studies using samples derived from reclaimed water concentrates. Only two studies—one in Denver (Lauer et al., 1990) and one in Tampa (CH$_2$M Hill, 1993)—have used live animal testing. While the Tampa study documented mutagenic activity in some samples through *in vitro* testing, no sample produced a significant positive response *in vivo*. Therefore, within the limits of the sensitivity of the *in vivo* tests that were used, the mutagenicity tests were not predictive.

In evaluating the safety of reclaimed water, additional uncertainties also exist in extrapolating the results from one location to another and from the present to the future at a given location. The confidence with which data can be extrapolated from one circumstance to another depends upon the completeness of the data and the representativeness of the samples.

In setting guidelines for evaluating the chemical safety of reclaimed water, it may be useful to consider the two general approaches used in safety testing of other products. The first approach is used for products whose chemical composition can be reasonably well defined. In general, such products are pure chemicals or relatively simple mixtures. For such products, very specific safety standards can be established.

A second approach, for products having poorly defined chemical composition, is to provide strict guidelines on production procedures. Standards on the chemical composition of such products can still apply, but they are supplemented by the guidelines and regulations regarding production techniques. For this type of situation, manufacturing may be approved on a plant-by-plant basis to ensure that production methods meet relevant guidelines.

Both approaches are used to regulate drinking water from non-reclaimed sources: maximum contaminant levels set explicit standards for specific chemicals, but treatment processes are also regulated. A similar combined approach can be used to regulate the toxicological safety of reclaimed water. That is, standards can be set specifying the types of toxicological testing that the water must undergo, but the processes for "manufacturing" the reclaimed water can and should also be strictly regulated.

EPIDEMIOLOGICAL METHODS FOR EVALUATING HEALTH RISKS

Human epidemiologic studies attempt to measure adverse health effects in a population exposed to a health hazard. A related but distinct activity is public health surveillance (described in Chapter 6), which collects information on morbidity and mortality but does not determine risk. Surveillance systems may reveal trends in disease morbidity and mortality but do not necessarily relate these trends to water quality. Determining whether an observed pattern of disease is associated with exposures to contaminants in drinking water requires a specifically designed epidemiologic study.

The use of epidemiologic methods to study health risks associated with drinking water has been recently reviewed by Craun et al. (1996) and Savitz and Moe (1997). Most epidemiologic investigations of drinking water and health have been conducted following waterborne disease outbreaks, when the investigators were primarily concerned with identifying the factors causing the particular outbreak and controlling the outbreak as quickly as possible. Epidemiologic studies designed to detect endemic waterborne disease or other health risks that may be associated with low levels of microbial or chemical contaminants in drinking water have proved more challenging.

In situations where there have been no obvious health problems or observed outbreaks and where endemic waterborne disease is low, the study population must be large enough to allow investigators to accurately measure and calculate whether a true difference exists between the disease patterns in the exposed population and the disease patterns in the unexposed population. In the study design phase, statistical-power calculations can be used to indicate the minimum risk that can be detected with a specific study population size. In the data-analysis phase, statistical-power calculations can also be used to interpret negative findings by providing information about the minimum risk that could have been detected given the size of the study population and the measured disease rates. Epidemiologic studies cannot prove that reclaimed water poses no risk; rather, negative findings can only show that the risk, if any, was less than what the study was capable of detecting.

Another critical issue in epidemiologic studies of reclaimed water is the problem of comparability. Any epidemiologic study requires comparisons of health outcomes or exposure experiences between different populations. In areas using or considering the use of reclaimed water, reclaimed water may be of better quality than other area water sources. If so, defining the appropriate "unexposed" or control group is difficult. The "unexposed" group may be a population in the same geographic area that consumes poor-quality water that is not reclaimed water, a population in a different geographic area that consumes high-quality water that is not available in the study area, or a population consuming bottled water. When considering the possible health risks associated with potable water reuse, one must also consider the risks associated with these other available water sources.

Basic Study Designs

Ecologic studies, also known as geographic studies, are often used in investigating the possible health effects of exposure to environmental contaminants. These studies are descriptive, are relatively inexpensive and easy to perform, and usually take advantage of existing data on mortality, morbidity, and demographics. Exposure and health outcome are characterized on an aggregate level. For example, initial studies of the potential health risks from chlorination by-products in drinking water

used ecologic designs in which exposure and cancer rates were characterized on a community level. Cancer mortality rates in communities with
chlorinated surface water supplies were compared to the rates in control
or "unexposed" communities served by ground water.

Ecologic studies are valuable for developing hypotheses that can be
tested in analytical epidemiologic studies. However, ecologic studies have
limitations. For instance, such studies do not take into account individual
risk factors (such as smoking or occupational exposure) or individual
water exposure patterns (such as the use of bottled water or the length of
residence in the study community). In addition, they assume that the
distribution of individual risk factors and behaviors will be relatively
equal in the exposed and control communities. In studies examining potential health effects with long latency periods (such as cancer), the results may be weakened by a mobile study population, since migration in
and out of the study area over time will reduce the sample size of the
truly exposed population.

In another type of study, the case-control study, the exposure histories of individuals with the disease of interest ("cases") are compared to
the exposure of individuals without the disease ("controls"). With this
design, cases and controls are queried directly about their residence history and water consumption habits, which provides greatly improved
estimates of risk associated with exposure to water from different sources.
Case-control studies allow the association between exposure and a single
disease or health outcome to be evaluated while controlling for individual
risk factors. However, case-control studies cannot prove that the exposure caused the disease or health outcome, because they do not provide
evidence that the exposure preceded the disease. Case-control studies are
useful in examining risk factors for specific health outcomes and are generally more efficient than cohort studies (described below), especially for
rare health outcomes, because they require fewer participants for adequate statistical power.

In a cohort study, also known as a longitudinal study, the disease
rates among a group of people who are exposed to the substance of interest (such as reclaimed water) are compared over time to the disease rates
in a group of people who are not exposed to the substance. Because this
design identifies the study population and determines exposure before
the development of disease, it can be used to determine the temporal
relationship between exposure to reclaimed water and the development
of various health outcomes. These studies are typically the most expensive epidemiologic designs, especially if they require long follow-up
times. Retrospective cohort studies cost less and take less time than prospective ones. Studies of occupational exposures have used retrospective
cohort designs in situations where historical exposure of a group of indi-

viduals ("cohort") was well characterized; health outcomes are then measured in the past or present. Although this study design has some limitations (such as the difficulty of collecting accurate historical exposure data), it has potential for future epidemiologic studies of possible health effects associated with potable reuse.

Each of these study designs has strengths and limitations, and no single study can conclusively establish the risks that may be associated with exposure to reclaimed water. Epidemiologists must examine a body of evidence collected from several studies and study designs and consider the following questions to determine whether the association between exposure and disease is a causal relationship:

1. Temporal association: Did the exposure of concern precede the development of disease?

2. Study precision and validity: Was the study well designed and conducted, and was the study population sufficiently large to detect meaningful differences in exposure and health outcomes?

3. Strength of association: Was the measure of association large enough to be a credible rather than a spurious result?

4. Consistency: Is there consistency of results across several studies with various designs and study conditions?

5. Specificity: Does the exposure result in a specific health outcome?

6. Biological plausibility: Is there scientific evidence from other fields (such as toxicology) to suggest that the exposure of concern results in the observed health effects?

7. Dose-response relationship: Was there evidence that increased exposure resulted in greater risk and/or more severe health outcomes?

8. Reversibility: Does removal of the exposure of interest result in a reduction of risk and/or disease?

The absence of one or more of the above conditions does not rule out causality. For example, some causal relationships can be weak and can result in multiple health outcomes. However, taken together, these criteria allow critical evaluation of specific epidemiologic findings in a broader context.

Exposure Assessment

Estimating actual exposure to microbial pathogens or chemical contaminants in drinking water is a difficult task. In general, populations who are served by a given water supply are considered to be "exposed." Yet there can be a significant range in the degree and mode of exposure, depending on individual water consumption habits, household treatment devices, and variabilities within the treatment and distribution system. In addition, inhalation and contact also serve as potential transmission routes for waterborne diseases and must be included in estimates of exposure (Savitz and Moe, 1997).

In epidemiologic practice, exposure to waterborne chemical or infectious agents is assessed by either (1) routinely measuring the actual microbial pathogen or chemical contaminant of concern or a proxy (such as a microbial indicator organism) in tap water or (2) estimating the presence, and possibly the amount, of the contaminant based on characteristics of the water source and treatment processes (such as a chlorinated surface water supply versus an untreated ground water supply). Individual consumption of or exposure to the water supply can then be determined by conducting interviews that collect information on lifetime residence history and self-reported water consumption habits (Savitz and Moe, 1997). Attempts to reconstruct the lifetime water exposure history of deceased cases by interviewing family members or using the last residence listed on death certificates should be avoided since these methods may misclassify exposure and introduce bias into the study (Craun et al., 1996).

Biomarkers of Exposure or Susceptibility

There has been increasing interest in the use of biomarkers (biochemical or molecular markers) of exposure or susceptibility in epidemiologic studies. For example, epidemiologists can use serological surveys to determine the prevalence of specific antibodies in a population and then compare the prevalence of specific infections, such as *Cryptosporidium*, between populations with different water sources in order to study the endemic risks of waterborne cryptosporidiosis (Craun et al., 1996). Such techniques may strengthen our ability to detect and classify exposure and disease (especially early stages of disease), clarify the steps between exposure and development of disease, and study the role of host factors that may account for variation in response. These techniques also may enhance the use of epidemiologic data to provide individual and group risk assessments (Schulte, 1993). Biomarkers could be particularly valuable for studying health effects of reclaimed water where there are a variety of possible exposures (microbial and chemical) and health outcomes of interest (infectious diseases, cancers).

Biomarkers in epidemiologic studies should be approached with caution since they can introduce confounding factors or bias (Pearce et al., 1995). For example, appropriate laboratory techniques must be chosen to measure the relevant antibodies, and the results must be interpreted carefully, since the presence of antibodies to a particular microorganism does not explain whether the infection was water related and may not indicate how long ago the infection occurred.

Sensitivities to chemical agents can also be highly variable. Genetic polymorphisms are proteins of the same general class that exist in different forms. These different forms have different functional properties. Individuals who express one form versus another can exhibit different sensitivities to chemical or microbial agents. As a consequence, molecular markers of genetic polymorphisms can be used to identify susceptible subpopulations and can enhance epidemiologic research by more accurately describing risk in specific groups, rather than assuming an average risk for the whole population. For instance, a biomarker that may be of relevance to health-effect studies of reclaimed water is the glutathione *S*-transferase theta 1 (GSTT1) gene. It has been hypothesized that polymorphisms in the GSTT1 gene may contribute to susceptibility to colon and rectal cancer. GSTT1 is involved in detoxification of a variety of halogenated organic compounds, including organic compounds present in chlorinated drinking water (bromoform, dibromochloromethane, dichlorobromomethane, and dichloromethane) (Casanova et al., 1997; Smith et al., 1995; Thier et al., 1993). Laboratory studies suggest that individuals with one or more intact copies of the GSTT1 gene could experience a different risk for colon and rectal cancer if exposed to chlorination by-products or other halogenated organic compounds in water.

Outcome Measurement

As discussed in Chapter 3, microbial agents in water have been associated with a diverse range of health problems. Waterborne enteric and aquatic microorganisms can cause acute and persistent gastroenteritis, dysentery, infectious hepatitis, febrile illness, meningitis, myocardiopathy, herpangina, encephalitis, hemolytic uremic syndrome, poliomyelitis, conjunctivitis, pharyngitis, ear infections, wound infections, dermatitis, Legionnaires' disease, Pontiac fever, stomach cancer, and various types of toxin poisoning. Chemical agents in water are known to have various acute and chronic toxic effects, and to cause reproductive problems, and they have been associated with cancer (bladder, liver, colon, rectal, kidney, esophagus, stomach, pancreas, and breast), depending on the type of agent, concentration, and duration and route of exposure.

Waterborne illnesses range from common, mild symptoms to rare, severe events, with results ranging from a successful natural immune response to death (see Chapter 3). Given this wide array of health effects and the variety of types and concentrations of contaminants that may be found in reclaimed water, it can be difficult to choose appropriate health outcomes for epidemiologic studies of reclaimed water. The choice of health outcomes to focus on for a particular study will affect the size of the study population and choice of study design. Some designs, such as prospective cohort and cross-sectional, are more amenable to the examination of frequent episodes of common illnesses, while other designs (case-control and retrospective cohort) are recommended for the study of rare, severe outcomes. Measuring both acute and chronic health outcomes in a single study is very challenging, because different measurement strategies may be required. Changes in disease incidence are a more sensitive outcome measure for environmental exposure than death, because (1) incidence would be expected to change earlier than mortality in response to an environmental exposure, (2) incidence is not affected by survival factors, and (3) incidence data are gathered from diagnostic information in patient medical records, which may be more informative than cause-of-death information from death certificates (Sloss et al., 1996).

The best method for measuring health outcomes in epidemiologic studies of health effects associated with water exposure depends on the study design and the health outcomes in question. Prospective studies of endemic waterborne infectious diseases have used personal health diaries to record episodes of gastrointestinal symptoms (Payment et al., 1991). Case-control studies of disinfection by-products and cancer have used medical records or death certificates (Craun, 1993). Ecologic studies of reclaimed water have used passive surveillance for infectious diseases and tumor registries and death certificates for evidence of chronic diseases (Sloss et al., 1996). Some outbreak investigations have used various forms of active surveillance, including lab-based active surveillance, telephone surveys, and monitoring of illness in institutions such as nursing homes and other "sentinel" or high-risk populations (MacKenzie et al., 1994). The quality of health outcome data collected by these diverse methods can vary, and data quality should be considered in the interpretation of results. Underdiagnosis and underreporting of disease and misclassification of health status may bias the results of an epidemiologic study.

Epidemiologic studies of water-associated health effects must consider several attributes of the exposure-disease relationships of interest. The timing of exposure relative to development and detection of disease varies considerably for different health outcomes. For most chemical agents suspected of causing cancer, risk arises from prolonged exposure, and there may be prolonged latency. Epidemiologic studies of these health outcomes must choose study populations who have had (or will have) sufficient lengths of exposure and follow-up time. For some cancers the latency time may be as long as 15 to 30 years. In contrast, chemical exposure that may affect reproductive health outcomes is of interest only during a brief interval early in pregnancy. Waterborne infectious diseases typically have short incubation periods and occur in response to a "slug" of microbial contamination in the water supply. For both chronic and acute diseases, the time course can vary depending on dose of the agent, prior exposure to the agent, and specific host factors (Savitz and Moe, 1997).

The ability to attribute a specific health outcome to a particular waterborne agent is related to the specificity of the exposure-disease relationship (Savitz and Moe, 1997). Diseases such as acute gastroenteritis and bladder cancer have many known and potential causes, some of which are completely unrelated to waterborne exposures. Of all reported cases of a particular disease or set of symptoms, only a subset can be potentially attributable to specific water exposures. For infectious diseases, clusters of cases due to a point source of contamination may be identified by molecular typing of clinical and environmental isolates. In contrast, it is extremely difficult to ascribe chronic diseases such as cancer to specific waterborne contaminants. When there are multiple transmission routes of the contaminant or multiple causes of the health outcome, the study must consider the effects of confounding factors (such as occupational exposures, smoking, and health behaviors) that can distort the relationship between water exposure and health outcome.

Sources of Bias and Error

When planning epidemiologic studies or reviewing the results of previous studies, one must examine whether the results could be affected by systematic errors in the sampling strategy or data collection procedures. Nonrandom error in a study that leads to distorted results is called "bias." Epidemiologists have described four major types of bias (Greenberg, 1993), and certain types of study designs are more prone to specific types of biases.

Usually, the first consideration is the potential for bias in the selection of the study subjects, or selection bias. The study must select subjects who are representative of the population of interest. Selection bias is a design issue and cannot be corrected in data analysis. In ecologic studies, the whole community is theoretically included in the study. However, certain segments of the population may be underrepresented because of the way health outcome data are collected. For example, health care facilities in poor neighborhoods may not have the resources to accurately diagnose and report specific health outcomes. When selecting control communities for ecologic studies, it is important for the control community to be similar to the test community in every way (demographics, income, education, etc.) except for exposure to the substance of interest (i.e., the water supply). In case-control studies, how cases and controls are selected is critical to the correct design of the study. Cases and controls must be selected without knowledge of their exposure. The strengths and limitations of various recruitment strategies need to be carefully considered. For example, choosing prevalent cases of cancer rather than incident cases will favor survivors; choosing controls from a hospital may select people with disease conditions that could also be influenced by environmental factors; and choosing community controls via telephone recruitment will favor those who have telephones and are available to answer the telephone (i.e., those who spend more time in the home). In cohort studies, the study population must be selected without knowledge of disease. Loss of certain members of the cohort during the follow-up period may result in bias if those who drop out of the cohort are unique in a certain way and become underrepresented in the final cohort. An example of such a unique group is families with young children; these families may find that they do not have time to participate in studies that require maintaining health diaries or responding to extensive interviews.

Information or misclassification bias can occur in the determination of exposure or of health outcome. This bias results from systematic errors in measuring either the exposure or the disease. In water studies, misclassification of exposure is a major concern. Information on an individual's exposure to a specific water supply may be based on information from death certificates or water utility bills. However, this measure only indicates exposure at a specific point in time that may not be the exposure time frame relevant to the disease of interest. For example, risks of bladder cancer may be strongly associated with the type of water supply an individual was exposed to 20 years previously and may not be related to that person's current water supply. When evaluating this type of bias, it is important to determine whether it is "differential" or "nondifferential." Differential misclassification occurs when either (1) exposure was measured differently for cases than for noncases or (2) health outcomes were measured differently for exposed than for unexposed people. Differential misclassification will result in either an overestimate or an underestimate of the measure of association between exposure and disease. Nondifferential misclassification occurs when either (1) exposure is measured equally incorrectly for both cases and controls or (2) health outcomes are determined equally incorrectly for the exposed and unexposed people. In these situations, the measure of association between exposure and disease is underestimated.

Recall bias can be a concern when data are gathered retrospectively and is most likely to occur in exposure assessment. For example, lifetime water histories may be recalled differently by cases than by controls, because ill people may be more likely to remember (and blame) specific exposures for their illness.

Confounding bias is a problem when the exposed and unexposed groups differ in the occurrence of some factor or factors that affect the development of the health outcome of interest. Common confounders are demographic characteristics such as age, gender, race, and occupation or behavioral characteristics such as smoking or alcohol abuse. For example, in an ecologic study, the study community may house a chemical manufacturing plant, and therefore it may have more individuals with occupational exposure to hazardous chemicals than the control community. This difference, rather than water supply, may account for higher cancer rates in the study community. The effect of confounding bias on the measure of association can be controlled in the study design either by restricting the study population to persons with a narrow range of a confounder (such as white, male, nonsmokers between the ages of 30 and 50) or by matching study groups based on confounders (such as matching cases and controls for age, gender, race); or confounding bias can be controlled in the analysis phase of the study by using techniques that account for the effect of the bias. All of these control methods rely on being able to identify and measure the relevant confounders for the health outcome of interest.

CONCLUSIONS AND RECOMMENDATIONS

A number of methodological issues make it difficult to definitively determine the public health risks of drinking reclaimed water.

Despite these uncertainties, any utility considering the implementation of a potable reuse project should estimate the increased risk from microbial and chemical contaminants in reclaimed water relative to those from other available sources of water.

Microbiological Methods and Risk Assessment

There is a lack of information nationwide on the levels of viral and protozoan pathogens in all waters and the efficacy of both conventional water treatment and wastewater treatment for water reclamation in reducing these levels. The Information Collection Rule, promulgated in 1996 by the EPA, is a first step toward providing some of the exposure data needed for more effective risk assessments, but additional steps are needed to improve methods for assessing risks posed by microbial pathogens in water reuse projects.

Potable reuse projects should consider using some of the newer analytical methods, such as biomolecular methods, as well as new in-

dicator microorganisms, such as *Clostridium perfringens* and the F-specific coliphage virus, to screen drinking water sources derived from treated wastewaters. The microbial methods currently used for detecting bacterial, viral, and protozoan pathogens in water all have limitations when used to detect pathogens in reclaimed water. Bacterial techniques do not account for viable, noncultivable bacteria, and the new techniques for assessing the viability of protozoan cysts or oocysts will require evaluation. Standard cell culture methods employing tests of cytopathic effects have been limited to the detection of well-known enteroviruses and do not account for many other enteric viruses that may be found in wastewater. New analytical methods for rapid measurement of health-related microbial contaminants, such as the polymerase chain reaction, are being developed. However, treatment performance and water quality goals have not been developed for these methods. In addition, several new indicator microorganisms are available, including *Clostridium perfringens* and the F-specific coliphage virus, and should be considered as alternatives to using the coliform bacteria as indicators for water quality. In particular, F-specific coliphage can be used in seeded studies to provide useful data on unit process removals.

The EPA should include data on the concentrations of waterborne pathogens in source water in the new Drinking Water National Contaminant Occurrence Data Base and should develop better data on reductions of waterborne pathogens by various levels of treatment. The lack of monitoring data for evaluating exposure remains the greatest single barrier to the development of risk assessment for microbial pathogens. Microbial risk assessment requires better estimates of exposure, which should be based on monitoring data, to identify the concentration of microbial pathogens in raw wastewater, wastewater treated with various processes, ambient water, and drinking water treated with various processes.

State officials, water utilities, and water research scientists should document survival rates of relevant protozoa in natural environments. Indirect potable reuse projects may rely on dilution in the environment and reduction by natural processes (i.e., die-off in ambient waters and removal by soil infiltration systems) to remove pathogens of all kinds; however, while reductions of bacteria and viruses have been well documented, the information on protozoa survival in ambient waters remains inadequate. More research is needed to fill that gap.

Risk estimates should consider the effects on sensitive populations and the potential for secondary spread of infectious disease within a community. This precaution is necessary to prevent pathogens from infecting sensitive populations (the elderly or very young, or those with suppressed immune systems) in whom mortality may be high and from whom diseases might spread to others.

The research community should conduct further studies to document the removal of pathogens of all types by membrane processes. Membrane systems, such as microfiltration, ultrafiltration, nanofiltration, and reverse osmosis, show the potential for nearly complete rejection of pathogens above certain size classes (in the case of the latter three processes, all size classes). More work is needed to demonstrate the suitability of these processes for potable reuse applications and to develop monitoring methods capable of continuously assessing process performance.

Chemical Risk Assessment

Analysis of reclaimed water is complicated by the fact that toxicological data on the wide variety of organic compounds in wastewater are much less complete than those for inorganic compounds. Thus, even if the organic material in the water could be analyzed completely, there would be no basis for assigning risks to most of the identifiable compounds present. This situation complicates the management of risks from chemical contaminants.

A conventional toxicological safety testing strategy developed in the food and drug industries uses both live animal (*in vivo*) and cell culture (*in vitro*) testing. While this approach has been used to develop risk assessments and regulations for recognized chemical contaminants in drinking water, there has been little experience in applying the strategy to determine health risks posed by the poorly characterized mixtures of organic chemicals in reclaimed water. The current state of knowledge in toxicology is too limited to make meaningful risk assessments based on *in vitro* data alone. So far, most toxicological studies of potable reuse have focused on bacterial and/or *in vitro* mammalian tests of genotoxicity of product waters rather than comprehensive testing on live animals. These bacterial or *in vitro* tests do not accurately evaluate the risks posed by the complex mixtures of contaminants in reclaimed wastewater.

Because of the uncertainty in the organic composition of reclaimed water, toxicological testing should be the primary component of chemical risk assessments of potable reuse systems. Attempting to ensure the safety of reclaimed water by analyzing only for known chemical contaminants, such as those regulated under the Safe Drinking Water Act, will not provide adequate protection of public health.

In waters where toxicological testing appears to be important for determining health risks, emphasis should be placed on live animal test systems capable of expressing a wide variety of toxicological ef-

fects. Chapter 5 presents a proposed system using fish in ambient waters.

Further, toxicological testing standards for reclaimed water should be supplemented by strict regulation of the processes for "manufacturing" the water. Regulators should review the processes for manufacturing the reclaimed water (that is, the treatment systems and environmental storage employed) on a plant-by-plant basis.

Epidemiological Methods

Several methodological challenges complicate epidemiologic investigations of the health effects of potable water reuse. These challenges include (1) obtaining accurate measures of individual or group exposure to the waterborne agent of interest, (2) selecting the appropriate health outcomes to monitor, (3) accurately measuring health outcomes through either a surveillance system or individual health records, (4) estimating the fraction of disease cases due to waterborne exposure, (5) selecting an appropriate comparison group, and (6) recruiting a study population large enough to detect a true effect.

Given these challenges, the results of epidemiologic studies should be interpreted with caution and a recognition of the potential for systematic and random error and potential biases. The strongest observational study design for establishing a cause-and-effect relationship between exposure to waterborne disease agents and disease occurrences is a cohort study, which compares the disease rates over time among individuals who are exposed to reclaimed water to disease rates among individuals who use a different water source.

REFERENCES

Asano, T., L. Y. C. Leong, M. G. Rigby, and R. H. Skaaji. 1992. Evaluation of the California wastewater reclamation criteria using enteric virus monitoring data. Water Science and Technology 26:1513-1524.

Ashby, J., and R. W. Tennant. 1988. Chemical structure, *Salmonella* mutagenicity and extent of carcinogenicity as indicators of genotoxic carcinogenesis among 222 chemicals tested in rodents by U.S. NCI/NTP. Mutation Research 204:17-115.

Atlas, R. M., G. Sayler, R. S. Burlage, and A. K. Bej. 1992. Molecular approaches for environmental monitoring of microorganisms. BioTechniques 12:706-717.

Bailey, N. T. J. 1975. The Mathematical Theory of Infectious Diseases and Its Applications. New York: Oxford University Press.

Bailey, N. T. J. 1986. Macro-modeling and prediction of epidemic spread at community level. Mathematical Modeling 7: 698-717.

Baron, R. C., F. D. Murphy, H. B. Greenberg, C. E. Davis, D. J. Bergman, G. W. Gary, J. M. Hughes, and L. B. Schonberher. 1982. Norwalk gastrointestinal illness: an outbreak associated with swimming in a recreational lake and secondary person to person transmission. American Journal of Epidemiology 115(2):163-172.

Bej, A. K., M. H. Mahbubani, R. Miller, J. L. DiCesare, L. Haff, and R. M. Atlas. 1990. Multiplex PCR amplification and immobilized capture probes for detection of bacterial pathogens and indicators in water. Molecular and Cellular Probes 4:353-365.

Benenson, A. S. (ed.) 1995. Control of Communicable Diseases Manual, 16 ed. Washington, D.C.: American Public Health Association.

Cabelli, V. J. 1997. *Clostridium perfringens* as a water quality indicator. Pp. 247-264 in Hoadley, A. W., and B. J Durka (eds.) Bacterial Indicators/Health Hazards Associated With Water. ASTM STP 635. Philadelphia: American Society for Testing and Materials.

Cabelli, V. J., A. P. Dufour, M. A. Levin, L. J. McCabe, and P. W. Haberman. 1979. Relationship of microbial indicators to health effects in marine bathing beaches. American Journal of Public Health 69(7):690-696.

Casanova, M., D. Bell, and H. Heck. 1997. Dichloromethane metabolism to formaldehyde (HCHO) and reaction of HCHO with nucleic acids in hepatocytes of rodents and humans with and without glutathione S-transferase T1 and M1 genes: adduct dosimetry and risk assessment. Submitted for publication.

CH_2M Hill. 1993. Tampa Water Resource Recovery Project. Tampa, Flo.: CH_2M Hill.

Clancy, J. L., W. D. Gollnitz, and Z. Tabib. 1994. Commercial labs: how accurate are they? Journal of the American Water Works Association 86:89-97.

Clausen, E. A., B. L. Green, and W. Litsky. 1977. Fecal streptococci: indicator of pollution. Pp. 247-264 in Hoadley, A. W., and B. J. Dutka (eds.) Bacterial Indicators/Health Hazards Associated With Water. ASTM STP 635. Philadelphia: American Society for Testing and Materials.

Cohen, J., and H. I. Shuval. 1973. Water, Air, and Soil Pollution 2:85-95. New York, N. Y.: Springer-Verlag.

Coin, L. 1966. Modern microbiology and virological aspects of water pollution. Pp. 1-10 in Jaag, O. (ed.) Advances in Water Pollution Research. London: Pergamon Press.

Craun, G. F. 1993. Epidemiology Studies of Water Disinfectants and Disinfection By-products. Pp. 277-301 in Craun, G. F. (ed.) Safety of Water Disinfection: Balancing Chemical and Microbial Risks. Washington, DC: ILSI Press

Craun, G. F., R. L. Calderon, and F. J. Frost. 1996. An introduction to epidemiology. Journal of the American Water Works Association 88:54-65.

Davies-Colley, R. J., R. J. Bell, and A. M. Donnison. 1994. Sunlight inactivation of enterococci and fecal coliforms in sewage effluent diluted in seawater. Applied Environmental Microbiology 60:2049-2058.

Deng, M. Y., S. P. Day, and D. O. Cliver. 1994. Detection of hepatitis A virus in environmental samples by antigen-capture PCR. Applied and Environmental Microbiology 60:1927-1933.

Douglas, G. R., D. H. Blakely, and D. B. Clayson. 1988. IPCEMC Working Paper No. 5: Genotoxicity tests as predictors of carcinogens: An analysis. Mutation Res. 196: 83-93.

Dufour, A. P. 1984. Bacterial indicators of recreational water quality. Canadian Journal of Public Health 75:49-56.

Eisenberg, J., E. Seto, A. W. Olivieri, and R. C. Spear. 1996. Quantifying water pathogen risk in an epidemiological framework. Risk Analysis 16(4):549-563.

Fujioka, R. S., and L. K. Shizumura. 1985. *Clostridium perfringens*, a reliable indicator of stream water quality. Journal of the Water Pollution Control Federation 57:986-992.

Fung, D. Y. C. 1994. Rapid methods and automation in food microbiology: a review. Food Reviews International 10(3): 357-375.

Garcia, L. S., T. C. Brewer, and A. Bruckner. 1987. Fluorescence detection of *Cryptosporidium* oocysts in human fecal specimens by using monoclonal antibodies. Journal of Clinical Microbiology 25:119-121.

Gerba, C. P., and J. B. Rose. 1990. Viruses in source and drinking water. Pp. 380-396 in McFeters, G. A. (ed.) Drinking Water Microbiology. New York: Springer-Verlag.

Gerba, C. P., J. B. Rose, and C. N. Haas. 1996. Sensitive populations: who is at greatest risk? International Journal of Food Microbiology 30(1-2):113-123.

Greenberg, R. S. 1993. Medical Epidemiology. Norwalk, Conn.: Appleton and Lange.

Griffin, P. M., and R. V. Tauxe. 1991. The epidemiology of infections caused by *Escherichia coli* 0157:h7, other enterohemorrhagic *E. coli* and the associated hemolytic uremic syndrome. Epidemiologic Reviews 13: 60-98.

Haas, C. N. 1983. Estimation of risk due to low doses of microorganisms: A comparison of alternative methodologies. American Journal of Epidemiology 118:573-582.

Haas, C. N., J. B. Rose, C. P. Gerba, and S. Regli. 1993. Risk assessment of viruses in drinking water. Risk Analysis 13:545-552.

Haas, C. N., C. Crockett, J. B. Rose, C. Gerba, and A. Fazil. 1996. Infectivity of *Cryptosporidium parvum* oocysts. Journal of the American Water Works Association 88(9):131-136.

Havelaar, A. H., M. Van Olphen, and Y. C. Drost. 1993. F-specific RNA bacteriophages are adequate model organisms for enteric viruses in fresh water. Applied and Environmental Microbiology 59:2956-2962.

Hurst, C. J., K. A. McClellan, and W. H. Benton. 1988. Comparison of cytopathogenicity, immunofluorescence and in situ DNA hybridization as methods for the detection of adenoviruses. Water Research 22:1547-1552.

Hurst, C. J., W. H. Benton, and R. E. Stetler. 1989. Detecting viruses in water. Journal of the American Water Works Association 9:71-80.

ILSI. 1996. Disinfection by-products in drinking water: critical issues in health effects research. Pp. 110-120 in Workshop Report. Chapel Hill, N.C.: October 23-25, 1995.

Jakubowski, W., S. Boutros, W. Faber, R. Fayer, W. Ghiorse, M. LeChevallier, J. Rose, S. Schaub, A. Singh, and M. Stewart. 1996. Environmental methods for *Cryptosporidium*. Journal of the American Water Works Association 88(9):107-121.

Johnson, D. W., N. J. Pieniazek, D. W. Griffin, L. Misener, and J. B. Rose. 1995. Development of a PCR protocol for sensitive detection of *Cryptosporidium* in water samples. Applied and Environmental Microbiology 61(11):3849-3855.

Kopecka, H., S. Dubrou, J. Prevot, J. Marechal, and J. M. Lopez-Pila. 1993. Detection of naturally occurring enteroviruses in waters by reverse transcription, polymerase chain reaction, and hybridization. Applied and Environmental Microbiology 59:1213-1219.

Lauer, W. C., F. J. Johns, G. W. Wolfe, B. A. Meyers, L. W. Condie, and J. F. Borzelleca. 1990. Comprehensive health effects testing program for Denver's potable reuse demonstration project. J. Toxicol. Environ. Health 30:305-321.

LeChevallier, M. W., and W. D. Norton. 1993. Treatment to address source water concerns: protozoa. Pp. 145-164 in Craun, G. F. (ed.) Safety of Water Disinfection: Balancing Chemical and Microbial Risks. Washington, D.C.: ILSI Press.

LeChevallier, M. W., and T. M. Trok. 1990. Comparison of the zinc sulfate and immunofluorescence techniques for detecting *Giardia* and *Cryptosporidium*. Journal of the American Water Works Association 82:75-82.

LeChevallier, M. W., W. D. Norton, and R. G. Lee. 1991a. *Giardia* and *Cryptosporidium* in filtered drinking water supplies. Applied and Environmental Microbiology 57:2617-2621.

LeChevallier, M. W., W. D. Norton, and R. G. Lee. 1991b. Occurrence of *Cryptosporidium* and *Giardia* spp in surface water supplies. Applied and Environmental Microbiology 57:2610-2616.

Linquist, H. A. D. 1997. Probes for specific detection of *Cryptosporidium parvum*. Wat. Res. 31(10):2668-2671.

MacKenzie, W. R., N. J. Hoxie, M. E. Proctor, M. S. Gradus, K. A. Blair, D. E. Peterson, J. J. Kazmierczak, D. G. Addiss, K. R. Fox, and J. B. Rose. 1994. A massive outbreak in Milwaukee of *Cryptosporidium* infection transmitted through the public water supply. New England Journal of Medicine 331:161-167.

MacKenzie, W. R., W. L. Schell, K. A. Blair, D. G. Addiss, D. E. Peterson, N. J. Hoxie, J. J. Kazmierczak, and J. P. Davis. 1995. Massive outbreak of waterborne *Cryptosporidium* infection in Milwaukee, Wisconsin: recurrence of illness and risk of secondary transmission. Clinical Infectious Diseases 21:57-62.

Mahbubani, M. H., A. K. Bej, M. Perlin, F. W. Schaefer, W. Jakubowski, and R. M. Atlas. 1991. Detection of *Giardia* cysts by using the polymerase chain reaction and distinguishing live from dead cysts. Applied and Environmental Microbiology 57:3456-3461.

Millard, P.S., K. F. Gensheimer, D. G. Addiss, D. M. Sosin, G. A. Beckett, A. Houck-Jankoski, and A. Hudson. 1994. An outbreak of cryptosporidiosis from fresh-pressed apple cider. Journal of the American Medical Association 272(20): 1592-1596.

Morens, D. M., R. M. Zweighaft, T. M. Vernon, G. W. Gary, J. J. Eslein, B. T. Wood, R. C. Holman, and R. Dolin. 1979. A waterborne outbreak of gastroenteritis with secondary person to person spread. Lancet May 5: 964-966.

National Institutes of Health (NIH). 1993. A chemical mixture of 25 groundwater contaminants. National Toxicology Program. NIH Report 93-3384. Research Triangle Park, NC: NIH.

National Research Council (NRC) 1977-1989. Drinking Water and Health, vols. 1-9. Washington, D.C.: National Academy Press.

National Research Council (NRC). 1982. Quality Criteria for Water Reuse. Washington, D.C.: National Academy Press.

National Research Council (NRC). 1983. Risk Assessment in the Federal Government: Managing the Process. Washington, D.C.: National Academy Press.

National Research Council (NRC). 1989. Drinking Water and Health: Selected Issues in Risk Assessment, Vol. 9. Washington, D.C.: National Academy Press.

National Research Council (NRC). 1994. Science and Judgment in Risk Assessment. Washington, D.C.: National Academy Press.

National Research Council (NRC). 1996. Use of Reclaimed Water and Sludge in Food Crop Production. Washington, D.C.: National Academy Press.

New York City's Advisory Panel on Waterborne Disease Assessment. 1994. Report of New York City's Advisory Panel on Waterborne Disease Assessment. October 7. The New York City Department of Environmental Protection.

Ongerth, J. E., and H. H. Stibbs. 1987. Identification of *Cryptosporidium* oocysts in river water. Applied and Environmental Microbiology 53: 672-679.

Parodi, S., M. Taning, P. Russo, M. Pala, D. Vecchio, G. Fassina, and L. Santi. 1983. Quantitative predictivity of the transformation in vitro assay compared with the Ames test. J. Toxicol. Environ. Health 12:483-510.

Paul, J. R., and J. D. Trask. 1947. The virus of poliomyelitis in stools and sewage. J. Am. Med. Assoc. 116:493-497.

Paul, J. H., J. B. Rose, J .Brown, E. A. Shinn, S. Miller, and S. Farrah. 1995. Viral tracer studies indicate contamination of marine waters by sewage disposal practices in Key Largo, Florida. Applied Environmental Microbiology 61:2230-2234.

Payment, P., and E. Franco. 1993. *Clostridium perfringens* and somatic coliphages as indicators of the efficiency of drinking water treatment for viruses and protozoan cysts. Applied and Environmental Microbiology 59:2418-2424.

Payment, P., and M. Trudel. 1985. Detection and health risk associated with low virus concentration in drinking water. Water Science and Technology 17:97-103.

Payment, P., M. Trudel, and R. Plante. 1985. Elimination of viruses and indicator bacteria at each step of treatment during preparation of drinking water at seven water treatment plants. Applied and Environmental Microbiology 49:1418-1428.

Payment, P., L. Richardson, J. Siemiatycki, R. Dewar, M. Edwardes, and E. Franco. 1991. A randomized trial to evaluate the risk of gastrointestinal disease due to consumption of drinking water meeting current microbiological standards. American Journal of Public Health 81(6):703-708.

Pearce, N., S de Sanjose, P. Boffetta, M. Kogevinas, R. Saracci, and D. Savitz. 1995. Limitations of biomarkers of exposure in cancer epidemiology. Epidemiology 6:190-194.

Pickering, L. K., D. G. Evans, H. L. DuPont, J. J. Vollet, III, and D. J. Evans, Jr. 1981. Diarrhea caused by shigella, rotavirus and giardia in day care centers: prospective study. Journal of Pediatrics 99(1): 51-56.

Powelson, D. K., and C. P. Gerba. 1995. Fate and transport of microorganisms in the vadose zone. Pp. 123-135 in Wilson, L. G., L. G. Everett, and S. J. Cullen (eds.) Handbook of Vadose Zone Characterization and Monitoring. Boca Raton, Flo.: Lewis Publishers.

Regli, S., J. B. Rose, C. N. Haas, and C. P. Gerba. 1991. Modeling the risk of *Giardia* and viruses in drinking water. Journal of the American Water Works Association 83:76-84.

Rose, J. B. 1997. Environmental ecology of *Cryptosporidium* and public health implications. Ann. Rev. Pub. Health 18: 135-161.

Rose, J. B., D. Kayed, M. S. Madore, C. P. Gerba, M. J. Arrowood, and C. R. Sterling. 1988a. Methods for the recovery of *Giardia* and *Cryptosporidium* from environmental waters and their comparative occurrence. In Wallis, P., and B. Hammond (eds.) Advances in Giardia Research. Calgary: University of Calgary Press.

Rose, J. B., H. Darbin, and C. P. Gerba. 1988b. Correlations of the protozoa, *Cryptosporidium* and *Giardia* with water quality variables in a watershed. Water Science and Technology 20:271-276.

Rose, J. B., L. K. Landeen, K. R. Riley, and C. P. Gerba. 1989. Evaluation of immunofluorescence techniques for detection of *Cryptosporidium* oocysts and *Giardia* cysts from environmental samples. Applied and Environmental Microbiology 55:3189-3195.

Rose, J. B., C. N. Haas, and S. Regli. 1991a. Risk assessment and control of waterborne giardiasis. American Journal of Public Health 81:709-713.

Rose, J. B., C. P. Gerba, and W. Jakubowski. 1991b. Survey of potable water supplies for *cryptosporidium* and *giardia*. Environmental Science and Technology 25:1393-1400.

Rose, J. B., L. J. Dickson, S. R. Farrah, and R. P. Carnahan. 1996. Removal of pathogenic and indicator microorganisms by a full scale water reclamation facility. Water Research 30(11):2785-2797.

Savitz, D., and C. L. Moe. 1997. Water: chlorinated hydrocarbons and infectious agents. Pp. 89-118 in Steenland, N. K., and D. A. Savitz (eds.) Topics in Environmental Epidemiology. New York: Oxford University Press.

Schulte, P. A. 1993. A conceptual and historical framework for molecular epidemiology. Pp 3-44 in Schulte, P., and F. R. Perera (eds.) Molecular Epidemiology: Principles and Practices. New York: Academic Press.

Singh, A., and G. A. McFeters. 1990. Injury of Enteropathogenic bacteria in drinking water. Pp. 368-379 in McFeters, G. A. (ed.) Drinking Water Microbiology. New York: Springer-Verlag.

Sinton, L. W., R. J. Davies-Colley, and R. G. Bell. 1994. Inactivation of enterococci and fecal coliforms from sewage and meatworks effluents in seawater chambers. Applied Environmental Microbiology 60:2040-2048.

Slifko, T. E., D. E. Friedman, J. B. Rose, and W. Jakubowski. 1997. An in vitro method for detecting infectious *Cryptosporidium* oocysts with cell culture. Applied and Environmental Microbiology 63(9): 3669-3675.

Sloss, E. M., S. A. Geschwind, D. F. McCaffrey, and B. R. Ritz. 1996. Groundwater Recharge with Reclaimed Water: An Epidemiologic Assessment in Los Angeles County, 1987-1991. Santa Monica, Calif.: RAND.

Smith, J. L., S. A. Palumbo, and I. Walls. 1993. Relationship between foodborne bacterial pathogens and the reactive athritides. Journal of Food Safety 13:209-236.

Smith, G., L. Stanley, E. Sim, R. C. Strange, and C. R. Wolf. 1995. Metabolic polymorphisms and cancer susceptibility. Cancer Survey 25:27-65.

Spika, J. S., J. E. Parsons, D. Nordenberg, J. D. Wells, R. A. Gunn, and P. A. Blake. 1986. Hemolytic uremic syndrome and diarrhea associated with *Escherichia coli* 0157:h7 in a day care center. Journal of Pediatrics 109: 287-291.

Stetler, R. E., R. L. Ward, and S. C. Waltrip. 1984. Enteric virus and indicator bacteria levels in a water treatment system modified to reduce trihalomethane production. Applied and Environmental Microbiology 47:319-324.

Stibbs, H. H., E. T. Riley, J. Stockard, J. Riggs, P. M. Wallis, and J. Issac-Renton. 1988. Immunofluorescence differentiation between various animal and human source Giardia cysts using monoclonal antibodies. Pp. 159-163 in Wallis, P., and B. Hammond (eds.) Advances in Giardia Research. Calgary: University of Calgary Press.

Sutter, T. E. 1995. Molecular and cellular approaches to extrapolation for risk assessment. Environ. Health Persp. 103:386-389.

Thier, R., J. B. Taylor, S. E. Pemble, W. G. Humphreys, M. Persmark, B. Ketterer, and F. P. Guengerich. 1993. Expression of mammalian glutathione S-transferase 5-5 in *Salmonella typhimurium* TA1535 leads to base-pair mutations upon exposure to dihalomethanes. Proc. Natl. Acad. Sci. 90:8576-80.

Tsai, Y. L., and B. H. Olson. 1991. Rapid method for direct extraction of DNA from soil and sediments. Applied and Environmental Microbiology 57 (4):1070-1074.

Tsai, Y. L., and B. H. Olson. 1992a. Detection of low numbers of bacterial cells in soils and sediments by polymerase chain reaction. Applied and Environmental Microbiology 58:754-757.

Tsai, Y. L., and B. H. Olson. 1992b. Rapid method for separation of bacterial DNA from humic substances in sediments for polymerase chain reaction. Applied and Environmental Microbiology 58:2292-2295.

U.S. Environmental Protection Agency (EPA). 1986. Ambient Water Quality Criteria for Bacteria. EPA 44015-84-002. Office of Regulations and Standards. Washington, D.C.: U.S. EPA.

U.S. Environmental Protection Agency (EPA). 1989a. Surface water treatment rule guidance manual for compliance with the filtration and disinfection requirements for public water systems using surface water sources. Federal Register 54(124). June 29.

U.S. Environmental Protection Agency (EPA). 1989b. Surface water treatment rule guidance manual for compliance with the filtration and disinfection requirements for public water systems using surface water sources. Federal Register 54(124) June 29.

U.S. Environmental Protection Agency (EPA). 1989c. National primary drinking water regulation; filtration and disinfection; turbidity; *Giardia lamlia*; viruses, *Legionella* and heterotophic bacteria. Federal Register 54:27486-27541.

U.S. Environmental Protection Agency (EPA). 1992. Guidelines for Water Reuse. EPA 625/R-92/004. Washington, D.C.: U.S. EPA.

U.S. Environmental Protection Agency (EPA). 1996. Proposed cancer risk assessment guidelines. EPA/600/P-92/003C. Washington, D.C.

5

Health-Effect Studies of Reuse Systems

While there is a general lack of toxicological and epidemiological data regarding potable reuse (see Chapter 4), a handful of such studies have specifically explored the public health implications of direct and indirect potable reuse. This chapter reviews six such health-effect studies conducted at operational or proposed planned potable reuse projects. Table 5-1 summarizes information from these studies. Most of the studies sought to analyze and compare the toxicological properties of reclaimed water to those of the current drinking water supply.

Windhoek is the only city in the world that has implemented direct potable reuse. The facility has operated since 1968 and has been the subject of epidemiological studies. In Denver, Colorado, direct reuse was studied extensively from about 1968 to 1992, but not adopted. In a related vein, a U.S. Army Corps of Engineers study was conducted in the early 1980s to assess the feasibility of using the wastewater-contaminated Potomac Estuary as a potential drinking water source for Washington, D.C. San Diego, California, and Tampa, Florida, have both conducted feasibility studies on adding reclaimed water to their surface water supplies and are moving toward implementation. Finally, California's Orange and Los Angeles counties, which have had operational indirect potable reuse systems in place for over 30 years, conducted a series of studies from 1975 to 1987 on the health effects of ground water recharge using reclaimed water.

TOXICOLOGY STUDIES

The studies conducted at the six projects varied from simple two-test studies to more comprehensive evaluations. The studies and their main findings are described briefly below; Table 5-1 provides further details.

Potomac Estuary Experimental Water Treatment Plant

In 1980, the U.S. Army Corps of Engineers began a two-year testing program of the Potomac Estuary Experimental Water Treatment Plant (EEWTP). Influent to the EEWTP was a 1:1 blend of estuary water and nitrified secondary effluent from the Blue Plains Wastewater Treatment Plant, which treats municipal wastewater from Washington, D.C. The blended water received further treatment by aeration, coagulation, flocculation, sedimentation, pre-disinfection, filtration, carbon adsorption, and post-disinfection.

Short-term *in vitro* tests (specifically, the Ames *Salmonella*/microsome test and a mammalian cell transformation test) were run on the EEWTP's blended influent, its effluent, and product water from three local conventional water treatment plants. For the toxicological parameters measured, the study found the EEWTP product water comparable to the finished waters from the local water treatment plants (James M. Montgomery, 1983). However, a National Research Council review panel (NRC, 1984) did not concur with this conclusion, because of the limited toxicological tests that were conducted.

Orange and Los Angeles Counties Health-Effect Study

The only toxicological study conducted to date on an operating indirect potable reuse project was performed as part of a five-year health-effect study, initiated in 1978, that evaluated possible effects resulting from surface spreading of reclaimed water in the Montebello Forebay area of Los Angeles County, California. Since inception of the potable reuse project in 1962, reclaimed water has been blended with local storm water and river water prior to percolation. At the time of the study, reclaimed water supplied about 16 percent of the total inflow to the ground water basin. Disinfected secondary effluent was used for recharge from 1962 to 1977, at which time dual-media filtration was added to the three wastewater treatment plants producing the product water. The toxicological study sought to detect, isolate, characterize, and if possible, trace

TABLE 5-1 Summary of Health-Effect Studies Evaluated

Project	Types of Water Studied	Health-Effect Data
Montebello Forebay, Los Angeles County, California (Nellor et al., 1984)	Disinfected filtered secondary effluent, storm runoff, and imported river water used for replenishment; also, recovered ground water	Toxicology testing: Ames *Salmonella* test and mammalian cell transformation assay. 10,000 to 20,000× organic concentrates used in Ames test, mammalian cell transformation assays, and subsequent chemical identification. The level of mutagenic activity (in decreasing order) was storm runoff > dry weather runoff > reclaimed water > ground water > imported water. No relation was observed between percent reclaimed water in wells and observed mutagenicity of residues isolated from wells Epidemiology: In the geographical comparison study, the population ingesting recovered water did not demonstrate any measurable adverse health effects. The household survey (women) found no elevated levels of specific illnesses or other differences in measures of general health
Denver Potable Water Reuse Demonstration Project (Lauer et al., 1990)	Advanced wastewater treatment (AWT) effluent (with ultrafiltration or reverse osmosis) and finished drinking water (current supply)	Toxicologic testing: 150 to 500× organic residue concentrates used in 2-year *in vivo* chronic/carcinogenicity study in rats and mice and reproductive/teratology study in rats. No treatment-related effects observed

Tampa Water Resource Recovery Project (CH$_2$M Hill, 1993; Pereira et al., undated)

AWT effluent (using GAC and ozone disinfection) and Hillsborough River water using ozone disinfection (current drinking water supply)

Toxicology testing: Up to 1000× organic concentrates used in Ames *Salmonella*, micronucleus, and sister chromatid exchange tests in three dose levels up to 1000× concentrates. No mutagenic activity was observed in any of the samples. *In vivo* testing included mouse skin initiation, strain A mouse lung adenoma, 90-day subchronic assay on mice and rats, developmental toxicity study on mice and rats, and reproductive study on mice. All tests were negative, except for some fetal toxicity exhibited in rats, but not mice, for the AWT sample

Total Resource Recovery Project, City of San Diego (Western Consortium for Public Health, 1996)

AWT effluent (reverse osmosis and GAC) and Miramar raw reservoir water (current drinking water supply)

Toxicology testing: 150-600× organic concentrates used in Ames *Salmonella* test, micronucleus, 6-thioguanine resistance, and mammalian cell transformation. The Ames test showed some mutagenic activity, but reclaimed water was less active than drinking water. The micronucleus test showed positive results only at the high (600×) doses for both treatments. *In vivo* fish biomonitoring (28-day bioaccumulation and swimming tests) showed no positive results

Epidemiology: baseline reproductive health and vital statistics

Neural tube defects study: No estimated health risk from chemicals identified based on use of reference doses and cancer potencies

Table continues on next page

TABLE 5-1 Continued

Project	Types of Water Studied	Health-Effect Data
Potomac Estuary Experimental Wastewater Treatment Plant (James M. Montgomery, Inc., 1983)	1:1 blend of estuary water and nitrified secondary effluent, AWT effluent (filtration and GAC), and finished drinking waters from three water treatment plants (current supplies)	Toxicology testing: 150× organic concentrates used in Ames *Salmonella* and mammalian cell transformation tests. Results showed low levels of mutagenic activity in the Ames test, with AWT water exhibiting less activity than finished drinking water. The cell transformation test showed a small number of positive samples with no difference between AWT water and finished drinking water
Windhoek, South Africa (Isaacson and Sayed, 1988)	AWT effluent (sand filtration, GAC)	Toxicological testing: Ames test, urease enzyme activity, and bacterial growth inhibition. *In vivo* tests include water flea lethality and fish biomonitoring (guppy breathing rhythm)
		Epidemiology: Study (1976-1983) of cases of diarrheal diseases, jaundice, and deaths. No relationships to drinking water source were found. Because of Windhoek's unique environment and demographics, these results cannot be extrapolated to other populations in industrial countries

NOTE: AWT = advanced wastewater treatment; GAC = granular activated carbon.

the origins of previously unidentified carcinogens in the ground water's replenishment sources and well waters.

The Ames test and *Salmonella* tester strains (TA98 and TA100) were used to screen for mutagenic organics in concentrates of reclaimed water (before it was spread), storm water, imported water, and unchlorinated and chlorinated ground water. While 13 of the 56 sample concentrates tested were free of mutagens, at least one mutagenic concentrate was found from each source. Nellor et al. (1984) reported that storm water and reclaimed water yielded the highest levels of mutagenicity, followed by well water and then imported waters. More than half of the mutagens observed appeared to derive from chlorination processes. The potency of the mutagenic responses did not appear to be related to the estimated percentages of reclaimed water at the various wells.

The study also compared the mutagenicity of samples of the ground water with mutagenic responses of known compounds and found that the ground water contained low concentrations of individual mutagens. While these tests (gas chromatography-mass spectroscopy (GC-MS) methods) found four identifiable Ames mutagens (fluoranthene, benzo-(*a*)pyrene, *N*-nitrosomorpholine, and *N*-nitropiperidine) in 6 of the 34 tested samples, Nellor et al. (1984) concluded that these compounds could not have caused the mutagenicity in all of the samples, because their frequency of occurrence, distribution in the fractions, and concentrations were not consistent with the bioassay results.

Further testing by chemical derivation techniques (including negative ion chemical ionization of GC-EIMS fractions and Ames assays of ground water and replenishment water before and after derivation) suggested that epoxides, organic halides, and two classes of electrophiles may have played a part in causing the observed mutagenicity. However, the results were not conclusive because the reactive components appeared only at part-per-trillion levels. The study identified neither the structures of these compounds nor their sources.

Since positive chemical identifications of specific mutagens could not be made and many of the estimated concentrations were very low, Nellor et al. concluded that the health significance of epoxides and organic halides at the levels found in the reclaimed waters remains in doubt. They stated that further characterization of the molecular structure and biological effects of the large numbers of apparently mutagenic halogenated organic compounds in the various waters would be necessary to confirm whether those materials pose any health risk.

Denver Potable Reuse Demonstration Project

The 3.8×10^3 m^3/day (1.0 million gallons per day (mgd)) Potable

Water Reuse Demonstration Plant in Denver, Colorado, began operations in 1984. The product water was treated by secondary treatment via biological oxidation, high-pH lime treatment, recarbonation, filtration, ultraviolet radiation, granular activated-carbon adsorption, reverse osmosis, air stripping, ozonation, and chloramination. A small ultrafiltration unit was evaluated as a possible replacement for reverse osmosis.

The health-effect studies, serving as a backup to analytical water quality monitoring, were implemented in the project's second year. They sought to evaluate the safety of the resulting product water compared to Denver's current drinking water, which comes from the Foothills Water Treatment Plant (Lauer, 1993). The studies incorporated acute toxicity testing, reproductive and teratogenic effects testing, subchronic toxicity testing, and chronic effects testing on animals.

The studies found that the quality of reclaimed water from the Denver Potable Reuse Demonstration Plant equaled or exceeded that of the existing drinking water supply and that it exceeded all federal and state standards for definable constituents (Lauer and Rogers, 1996). In addition, a two-year chronic toxicity/carcinogenicity study, which gave concentrated doses of organic samples to more than 1500 rats over two generations, revealed no toxicologic, carcinogenic, reproductive, or developmental effects.

San Diego Total Resources Recovery Project

The City of San Diego, California, imports virtually all of its water from other parts of the state, and current supplies are projected to be insufficient to meet future demands. San Diego investigated indirect potable reuse as one measure to help alleviate water shortages in the future. The San Diego Total Resources Recovery Project developed a series of pilot treatment facilities that included primary treatment by rotary disk filter, a 1.9×10^3 m^3/day (0.5 mgd) water hyacinth secondary treatment facility, and 190 m^3/day (0.05 mgd) advanced wastewater treatment (AWT) facilities. The AWT treatment train included coagulant addition, filtration, disinfection by ultraviolet radiation, reverse osmosis, air stripping via an aeration tower, and granular activated-carbon adsorption.

A health-effect study (Western Consortium for Public Health, 1992) compared the reclaimed water quality and its health risks to those of the city's current raw water supply from the Miramar reservoir. The study used four types of bioassay systems to evaluate genetic toxicity and potential cancer-causing effects. The tests used included the Ames test, the micronucleus test, the 6-thioguanine resistance assay, and the mammalian cell transformation assay. Forty-eight water samples collected between February 1988 and June 1990 were concentrated and then used for these various bioassay systems. In addition to the total concentrate, many of the fractions and a few of the subfractions were also tested.

The Ames test, which sought to measure hereditable genetic alterations in special strains of *Salmonella*, was performed on 23 samples of reclaimed water and 25 samples of Miramar water. It found weak but statistically significant mutagenic activity in a number of the samples from both waters. The Miramar water samples exhibited more mutagenic activity than those from the reclaimed water.

The mouse micronucleus assay, which is a short-term assay that assesses genetic damage to the bone marrow of mice exposed to concentrates, was run on seven samples of both waters. It revealed no statistically significant effect in the majority of samples from either water source until the doses of concentrates were raised to near-lethal levels, at which point a possible trend toward increased genetic damage was observed with both reclaimed water and Miramar water. An approximately threefold increase in micronuclei frequency was observed in the two high doses of reclaimed water samples, but was not confirmed by follow-up experiments.

The 6-thioguanine resistance assay measures mutagenic inactivation of a certain gene (known as HGPRT) in a cell line established from hamster ovaries. It was run repeatedly on one sample each of whole concentrate from reclaimed water and Miramar water and five fractions, but showed no apparent mutagenic effect.

Finally, the mammalian cell transformation assay, which measures the ability of chemicals to induce changes in a certain strain of cells (called C43H10T1/2 cells) that are injected into immunosuppressed mice, was run several times on one sample collected from each water. The Miramar sample produced a strong positive response that appeared to be dose related. However, the response was not observed in two other samples. As a result, the authors suggested that this single positive test was not significant.

In sum, the study found that organic extracts from both reclaimed water and Miramar water sources did exhibit some genotoxic activity, primarily in the Ames test and to a lesser extent in the other bioassays. The activity, however, was stronger in the Miramar water than the reclaimed water. The study's authors could or did not identify the reason behind the higher activity in Miramar water; however, they speculated that the greater activity exhibited by Miramar water may reflect differences in composition of original source waters, including chlorination of Miramar water. The report concludes that, based on short-term bioassay results of organic extracts of both reclaimed water and Miramar water, reclaimed water is unlikely to be more genotoxic or mutagenic than the current raw water supply source for San Diego.

Tampa Water Resource Recovery Project

In the 1980s, the City of Tampa, Florida, evaluated the acceptability of using reclaimed water to augment Tampa's current water supply from the Hillsborough River. An AWT pilot treatment plant was completed in 1986 and operated from January 1987 through June 1989. Toxicological testing of concentrates was completed in August 1992.

Influent to the AWT pilot facility consisted of treated wastewater from the Hookers Point Wastewater Treatment Facility, which provided secondary treatment, filtration, and denitrification. The pilot plant influent was withdrawn prior to disinfection to reduce the concentration of chlorinated organic compounds. The project evaluated four different unit process trains. All of the process trains included preaeration, high-pH lime treatment, recarbonation, gravity filtration, and disinfection with ozone. Three of the trains also included organic removal processes that differed only in selection of the unit process added between gravity filtration and disinfection; one train added granular activated carbon (GAC), one train added reverse osmosis, and one train added ultrafiltration. The pilot plant was originally operated using chlorine as the disinfectant, but results of Ames testing indicated that ozone-disinfected product waters were less mutagenic than chlorine-disinfected product waters. For similar reasons, the treatment train using GAC was selected for toxicological testing based on preliminary screening using the Ames assay.

The Hillsborough River water was disinfected with ozone prior to analysis to make it more analogous to the AWT pilot plant product water. Concentrated extracts from both the Hillsborough River water and the reclaimed water were used to create doses for toxicological testing at up to 1000 times the potential human exposure of a 70 kg (154-pound) person consuming 2 liters of water per day. Toxicological tests evaluated mutagenicity, genotoxicity, subchronic toxicity, reproductive effects, and teratogenicity.

Four concentrate samples from each of the two waters were tested extensively for mutagenic activity. In all, eight toxicological studies were conducted on the reclaimed water and ozone-disinfected Hillsborough River reference water. These tests evaluated potential genotoxic effects (Ames and sister chromatid assays), carcinogenicity (strain A lung adenoma and SENCAR mice initiation-promotion studies), fetotoxicity (teratology in rats and reproductive effects in mice), and subchronic toxicity (90-day gavage studies in mice and rats). The results were reported to be uniformly negative for the product water from the AWT pilot plant (Hemmer et al., 1994).

EVALUATION OF TOXICOLOGY STUDIES

Do we have sufficient data to indicate that reclaimed wastewater can be reliably used as a source of drinking water? To answer this question, the results of the toxicological studies described above must be examined against the "decision logic" for which the assay systems they used were designed. Important differences exist among the available toxicological studies regarding the intent and extent of toxicological testing done. The studies fall into three categories:

1. screening and identification studies,
2. surveys of mutagenic activity, and
3. integrated toxicological testing.

While several of the studies fit more than one of these categories, it is important to highlight the differences in philosophical approach that these categories represent.

Screening and Identification Studies

As described in more detail in Chapter 4, screening and identification studies seek to identify specific chemicals in the water sample that could present a hazard to health. This approach is an alternative to exhaustive chemical analyses. The approach does not attempt to measure risk or to be a comprehensive test of potential health effects. Rather, it seeks to identify compounds that could pose major problems at low doses (i.e., potential genotoxic carcinogens). Such an approach assumes that the risks posed by any compounds so identified will receive additional study, either through the literature or by more complete characterization of their toxic effects in systems that would provide an acceptable basis for estimating risk.

This use of bioassays for screening is well accepted in the scientific community. Bioassays have been used in most of the world's extensive studies of drinking water, leading to the identification of numerous highly mutagenic chemicals produced in the chlorination of drinking water (Bull and Kopfler, 1991). These chemicals are now receiving the toxicological evaluations they warrant (Bull et al., 1995). The higher-level toxicological evaluations indicate that these chemicals are less important as potential carcinogens than would be anticipated from their mutagenic potency (ILSI, 1995). They present no greater a carcinogenic risk than do several nonmutagenic compounds, such as dichloroacetate, that are produced at much higher concentrations in chlorinated drinking water (ILSI, 1995).

The water reclamation projects reviewed here very rarely took the approach of screening with bioassays and then applying more detailed toxicological evaluations. In fact, this approach was applied only to one aspect of the health-effect study conducted by Orange and Los Angeles Counties (OLAC) (Baird et al., 1980, 1987; Jacks et al., 1983). The bulk of this OLAC work was performed using the *Salmonella*/microsome assay. The first efforts were directed at simply identifying the relative mutagenic activity of chemicals that could be isolated from differing source waters. Certain of the more mutagenic waters were then fractionated and studied in more detail. While the approach met with some modest success, the mutagenic activity of the fractions was greater than could be accounted for by known mutagens. Part of the discrepancy was probably caused by the reliance on gas chromatography-mass spectrometry (GC-MS) as the analytical tool, because most of the chemicals in water, including some that are potent mutagens, cannot be measured with GC-MS.

The OLAC study also attempted to use derivatization methods (which use a second chemical to react with and thus detect the target chemical) to detect certain chemical targets (Baird et al., 1987; Jacks et al., 1983). While this test produced some significant modifications of mutagenic activity in some water samples, it revealed no consistent pattern of mutagen reduction. This suggests that the chemicals present in different waters may be of different characters.

It is of interest to contrast the inconclusive nature of the OLAC study's results with the progress that has been made in drinking water chlorination. The initial studies of mutagenicity in drinking waters around the world first traced mutagenicity to the process of chlorination; the major mutagenic activity was found to be associated with a very potent mutagen referred to as MX (Meier et al., 1987), which is produced in chlorination. The whole-animal carcinogenesis testing of this chemical recently ended and concluded that MX induces tumors (Tuomisto et al., 1995). However, a recent workshop (ILSI, 1995) focused on the toxicology data for MX and concluded that MX is probably not a major contributor to cancer via drinking water because it is found at extremely low levels in chlorinated drinking waters.

Surveys of Mutagenic Activity

Other studies used the *Salmonella*/microsome assay and other tests to simply compare reclaimed wastewater to other drinking water sources.

The OLAC, Tampa, San Diego, Windhoek, and Potomac River studies included this type of effort.

The OLAC study found that residues of organic chemicals prepared by a high volume concentrator with samples taken from wastewater, storm water runoff, and chlorinated and unchlorinated wells had only minor differences in the mutagenic activities that could be isolated. On the other hand, imported water (state project water and Colorado River water) appeared to be consistently less mutagenic (Nellor et al., 1984).

One surprising finding was that the chlorinated wells were, if anything, lower in mutagenic activity than unchlorinated wells. However, the well waters reacted differently than do most chlorinated waters to the addition of a chemical called S-9, which in most waters sharply reduces mutagenic activity (Cheh et al., 1980; Meier et al., 1983), but which produced only modest decreases in the chlorinated OLAC waters. This difference suggests that the nature of the mutagenic activity found in the OLAC well water samples is atypical of that likely to be found in chlorinated surface water supplies in the United States. (It actually is more reminiscent of the types of activity that have been associated with the industrially polluted Rhine River in Europe in a series of studies by Dutch workers (Kool et al., 1982).)

The Tampa project also used the *Salmonella*/microsome assay to compare mutagenic activity of reclaimed water to Hillsborough River water (the current water supply). However, it added some potentially more meaningful tests by measuring clastogenic activity of samples *in vivo*. These tests were conducted on splenocytes that were isolated from mice treated with concentrated organic material from different waters. No positive results were found in ozone-disinfected GAC-treated product water or the Hillsborough River water. None of the samples exhibited increased frequencies of sister chromatid exchange or induction of micronuclei in splenocytes isolated from the treated mice. However, cytotoxicity occurred in the Hillsborough River water sample, which limited the amount of the concentrate that could be evaluated. This is an unusual result because the dosing was made *in vivo*, and cytotoxicity of this sort is more often associated with *in vitro* experiments. This result suggests that organic compounds concentrated to 300 times the normal dose from Hillsborough River water may have some effects on splenic function.

In the Tampa project, positive mutagenic activity was associated with both the river water and the AWT effluents after filtration, ultrafiltration, and reverse osmosis. In most of these cases the addition of chlorine tended to increase mutagenic activity in the Ames test, whereas ozonation produced inconsistent effects.

The interpretation of the results of the Tampa study is subject to the same ambiguity identified with other studies that have depended primarily on *in vitro* test systems. While screening assays showed some mutagenic activity, there was no attempt to collect further data that would allow a comparison of the relative actual risks of disease posed by these two water sources.

The Potomac River study used *in vitro* testing procedures. Here the Ames test detected mutagenicity in both the blended influent and the finished drinking water at about the same levels and with the same frequency (approximately 50 percent of the time) of positive results. The effluent tested positive only about 10 percent of the time.

The Potomac River study, using cell transformation assays employing 10T1/2 cells, obtained some positive results, but it is difficult to analyze data because negative controls were not included in the data set provided.

When *in vitro* mutagenicity testing is used alone, no clear meaning can be assigned to the results. When the test is used at different stages in the treatment process, a demonstrated reduction of mutagenic activity might be meaningful because it would represent simple removal of activity that was present. However, if reactive chemicals (e.g., disinfectants) modify mutagenic activity, it is impossible to determine whether this modification is meaningful in terms of health risk, because mutagenic activity alone does not imply carcinogenic risk to humans. A more accurate determination of health risk requires test systems that can more directly measure a more complete range of health hazards and define dose-response relationships that link degrees of risk to levels of exposure. Such information cannot be derived from data generated by *in vitro* testing alone.

A more serious drawback of relying only on mutagenic testing is that such testing provides no "toxicological characterization" of potential hazards. As pointed out in Chapter 4, negative results in such tests do not guarantee safety. This uncertainty has three sources. First, many adverse health effects of chemicals do not require a chemically induced mutation. Second, research has shown virtually no correlation between the genotoxic potency and the carcinogenic potency of chemicals. Third, subsequent work has shown that the collective false negative rate for carcinogenic chemicals in mutagenesis assays is much higher than initially believed (Ashby and Tennant, 1988), leaving more carcinogenic chemicals undetected than previously thought.

Integrated Studies of Health Effects of Reclaimed Wastewater

Only the Denver and Tampa studies addressed a broad range of toxicological concerns. The actual approaches in the two studies differed

Checking the ozone disinfection unit at San Diego's Aqua III water reclamation facility. Photo by Joe Klein.

somewhat, though both contained most of the elements recommended by the 1982 National Research Council report *Quality Criteria for Water Reuse* (NRC, 1982). For instance, both studies minimized use of *in vitro* screening techniques, as the 1982 NRC report suggested, focusing on tests on experimental animals instead; the Denver project did not use *in vitro* testing at all.

The Denver study used organic concentrates from reclaimed water and the finished water for Denver's drinking water treatment plant to produce 150x and 500x doses. The samples were administered to whole animals. The concentrates were fortified with a limited number of volatile organic compounds presumed to have been lost during the concentration procedure. The comparative testing included a chronic toxicity test combined with the equivalent of a two-year carcinogenicity bioassay in both rats and mice. This was supplemented by a study of reproductive toxicity in Sprague-Dawley rats in which the offspring were systematically evaluated for birth defects through two generations. The study did not include measures of mutagenic activity, either *in vitro* or *in vivo*.

These *in vivo* studies identified no treatment-related effects. This suggests that no adverse health effects can be anticipated from the potable reuse of Denver's wastewater. There was no effect in the animals at a dose level up to 500 times the concentration that humans would be exposed to. These findings of course apply to a specific point in time and the use of a particular treatment train, as derived from concentrated samples that may not be representative of the material in the source water. Despite these qualifications, there is reasonable certainty about the safety of use of this reclaimed water as a source of drinking water, because its chemical constituents were tested in systems commonly accepted as a basis for establishing the safety of a product.

Tampa's project on reclaimed water was similar in overall approach, but substituted some shorter-term *in vivo* carcinogenesis experiments for the two-year bioassay. These carcinogenicity tests were lung tumor induction in strain A/J mice and the mouse skin initiation/promotion assay in SENCAR mice. The use of these tests in place of two-year bioassays in two species increases the sensitivity of the test to some agents but tends to narrow the range of carcinogens that can be detected. Still, the use of an intact animal allows testing of a broader range of toxicity than do *in vitro* tests. Toxicity was also assessed with a reproductive study conducted in mice and with a teratogenesis assay conducted in rats. None of these tests produced positive findings of long-term effects despite using concentration doses of 100x, 300x, and 1000x. Using the same type of logic and qualifications that were applied to the Denver study, one could say with some confidence that this study indicates that reclaimed water carries no significant impact on health with a nominal margin of exposure of 1000.

In spite of these excellent efforts at conducting well-conceived, long-term toxicological studies, there are technical problems with the testing of organic concentrates. The preparation of representative concentrates of organics in water is not simple. While recoveries of organic material can approach 70 percent in some systems (Jenkins et al., 1983), they are often significantly lower in many studies. In the studies reviewed here, recovery levels were not always reported. Tampa reported 20.9 percent recovery from ozone-disinfected Hillsborough River water and 9.1 percent from ozone-disinfected GAC product water (CH$_2$M Hill, 1993). Recovery in various OLAC studies ranged from less than 50 percent (Baird et al., 1980) to 70 percent (Jenkins et al., 1983).

A second issue is the degradation of samples over time. While some effort was made to stabilize the samples, it is impossible to know that no changes occurred that could have affected the result. In a situation where major portions of the material cannot be identified, obtaining objective measures of stability is difficult.

A third commonly voiced criticism of using concentrates is that it is impossible to know whether the concentration procedure itself produces or destroys some products by accelerating chemical reactions. Each of these criticisms is relevant to the confidence of negative results.

Other criticisms deal with the completeness of the testing. While the tests applied had elements of currently accepted protocols for assessing the safety of commercial products (FDA, 1982; U.S. EPA, 1979), pragmatic concerns forced some potentially significant departures from conventional practice. For example, conventional practice in safety testing dictates that materials should be tested at the maximum tolerated dose (MTD). But in the Denver and Tampa studies, the cost of concentrate preparation limited the amount of concentrate that could be prepared, and the MTD was not approximated. The MTD may not have been approached even if the concentration factors had been increased by another order of magnitude. Nevertheless, in the opinion of the committee, these two studies approach the practical limit of the type of study that could be performed using organic concentrates of reclaimed wastewater.

An issue in any retrospective evaluation such as this is that public health concerns evolve over time. The focus of safety testing in the 1990s has moved beyond where it was in the 1970s and 1980s when most of these studies were performed. For example, there is now considerable concern about chemicals loosely referred to as endocrine disrupters (Kavlock et al., 1996). Much of the controversy over this type of chemical concerns estrogen-like chemicals such as dioxin and polychlorinated biphenyls (PCBs) and their potential relationships with diseases like breast cancer. The concern has recently broadened to include chemicals, such as alkylphenol ethoxylate, that have produced apparent estrogen effects in fish (see Chapter 2). Endocrine disrupters are specifically identified for evaluation of health impacts in the latest reauthorization of the Safe Drinking Water Act. A potentially important issue for potable use of reclaimed water is that chemicals producing endocrine disruption have been associated with municipal wastewater effluents (Sumpter, 1995).

Implications for Future Safety Testing

The Denver and Tampa studies found no signs of significant adverse effects of consumption of reclaimed water. However, two sets of data drawn from two discrete points in time and conducted at a pilot plant level of effort provide a very limited database from which to extrapolate to other locations and times.

If these data are inadequate, what more should be done? Clearly the approach taken in the Tampa and Denver studies does more to establish safety than other studies have. However, there are several reasons to believe that such testing will always be less than satisfactory.

One critical problem is with the preparation of the water sample to be tested. As long as the sample can be considered less than fully representative of the chemical constituents in the source water, the testing can be criticized as incomplete. As discussed in Chapter 4, organics are concentrated both to increase the effective dose for testing purposes and to separate inorganics that might dehydrate the test organisms. But the processes that concentrate the organics may create reactions that remove or add chemical compounds, thus changing the mixture of chemicals. So far we have no reliable way to verify how well a sample represents the water from which it is derived. As explained in Chapter 2, certain chemical characteristics can be used to describe the nature of the organic chemicals in water. It is possible that a confirmatory procedure could be developed that would (1) verify the consistency of the chemical characteristics of samples produced and (2) verify that the process of concentration did not cause chemical components of the mixture to react and change. Developing such a procedure would require a very significant research effort.

Another major issue is the expense of completing an adequate safety evaluation. Cost estimates should consider not only the investment for original testing at the pilot stage but also for ongoing measures to monitor and ensure the safety of the product water over time. Such ongoing efforts must address not only potential changes in the quality of the water but also changing priorities regarding what health risks should be addressed.

These difficulties suggest that alternative strategies for testing reclaimed water must be sought. One option is to employ conventional safety testing protocols used to evaluate new food additives and drugs. However, there are problems with this approach. One is that the decision logic used in conventional safety testing calls for testing at or approaching the MTD. The lack of substantive effects in the Denver and Tampa studies at 500- to 1000-fold concentration factors suggests that testing at the MTD is impractical, if not impossible.

A final problem is the timing of results. For potable reuse projects, continuous toxicity testing is desirable to provide project operations with an additional "warning system" in the event of unanticipated changes in product water quality. Conventional methods of toxicity testing do not allow for such continuous monitoring and the production of rapid results.

Another problem in applying the logic of safety testing to reclaimed water is that, unlike product developers, water utilities cannot simply drop their product lines. If a commercial product is shown to be mutagenic by simple *in vitro* tests, the producer can avoid the costs of further testing by terminating its production. This is not a reasonable option for drinking water, which always requires further testing. For example, most drinking waters in the United States would show a mutagenic response associated with disinfection (Cheh et al., 1980; Meier et al., 1987) if so tested; yet this does not mean the water is unsafe, since mutagenesis does not necessarily indicate carcinogenicity or other health threats. And in view of the well-established contribution that disinfection makes to public health, it would be foolish to discard either the water or the disinfection process simply because disinfection increases mutagenic activity. Thus, the question should not be whether a chemical is mutagenic in a bacterial system but whether it presents a carcinogenic hazard to humans and at what levels of exposure. And determining carcinogenicity requires tests of intact animals regardless of whether the *in vitro* mutagenesis test is positive or negative.

Final problems with trying to apply conventional safety testing to reclaimed water are timing of the results and determining what action is required if a positive response is detected in live animals. For example, what should be done if a chronic rodent study (which typically takes two years to run and another year to analyze) finds a marginally significant increase in the incidence of tumor-bearing animals exposed to the test water? It is highly unlikely that a specific chemical agent could be identified within a reasonable time frame. In this situation, one is caught in the dilemma of deciding whether this test (1) was a statistical fluke, (2) reflected some transient changes in water quality that occurred during the sample, or (3) represented a true hazard. The only way to answer the question might be to rerun the study. Consistent results from a second study would be clear cause for concern. But what if the second test was negative? Would a tiebreaker be needed? While the testing scheme proposed by the NRC (1982) was the proper approach to be taken from a toxicologist's point of view, the approach is exceedingly difficult to implement for testing reclaimed water projects. It also does not provide results rapidly enough to respond to important changes in water quality.

Thus, a critical difference between "new product" testing and the testing of reclaimed waters is the timing of the testing. In the case of new products, the system is designed to prevent the introduction of dangerous chemicals. Testing is often done before anyone is exposed to the chemical. Clear positive evidence of adverse health effects, even at doses substantially greater than would be anticipated in the environment, can prevent the product's introduction into commerce. On the other hand, once the product has been established as safe by appropriate testing, fairly straightforward manufacturing practice and quality control procedures should ensure continued safety of the product into the future. In the case of drinking reclaimed water, on the other hand, it is possible (although not highly likely) that a significant but unknown chemical hazard could be introduced into wastewater in such a subtle way that it may not be detected in time with conventional toxicity tests.

Potential New Approaches for Judging the Safety of New Water Sources

Potable reuse projects need a new approach to toxicity testing. Future toxicological characterizations of wastewaters intended for potable reuse or water derived for potable use from wastewaters should focus specifically on data needed for risk assessment. Because the substance being tested is essentially unknown, it is important that whole animals be used for testing to allow the concurrent evaluation of multiple end points. If only *in vitro* tests are conducted, the test system becomes a potentially large collection of independent tests that frequently cannot be integrated into a realistic estimate of human health risk. Also essential is a toxicity testing system that can allow continuous monitoring and produce timely results.

One toxicity testing system that has been the subject of increasing research since the early 1980s and that may meet the needs of potable reuse systems uses fish as the subjects of testing (see, for example, Anders et al., 1984; Bunton, 1996; Calabrese et al., 1992; Courtney and Couch, 1984; Hatanaka et al., 1982; Hawkins et al., 1984, 1985; Krause et al., 1997; Lopez and De Angelo, 1997; Sato et al., 1992; Walker et al., 1985). Some of this research has focused specifically on the use of fish for assessing the carcinogenicity of drinking water contaminants. Further research would be needed to develop fish-based toxicity testing systems for potable reuse projects, but a sufficient body of research exists to begin trying such an approach.

Using a fish testing system departs from previous recommendations of the NRC (1982) in four general ways:

1. The baseline screening test would use whole-animal testing rather than screening tests based largely on *in vitro* assays. While use of a mammalian species would be ideal because of the long experience in using such animals as surrogates for humans, using one or more fish species provides important practical advantages, such as cost and timeliness.

2. The baseline screening tests could be conducted using water samples at ambient concentrations. The uncertain and expensive use of concentrates could be abandoned.

3. Research efforts would need to be undertaken to identify and understand the qualitative and quantitative relationships of responses in fish test species to adverse health effects in humans.

4. *In vitro* short-term testing could be confined to qualitative evaluations of particular toxicological effects found in the product water in order to identify potential contaminants and quickly guide remedial actions. Such studies would probably have to use concentration techniques to increase the sensitivity of *in vitro* tests.

Use of Fish as the Baseline Test

Using fish to test the quality of wastewaters presents several advantages:

1. Exposure to chemicals in the water is continuous and does not require specialized procedures (such as preparing concentrates for frequent administration to rodents by stomach tube).

2. Large numbers of fish are much simpler and less costly to handle than large numbers of rodents.

3. Considerable current research and research over the last two decades has focused on certain toxicological end points in fish and on the similarities and differences in responses of fish, mammalian test species, and humans.

The high exposure of fish and the relative ease of maintaining large numbers of them offset some of the losses in sensitivity resulting from not using concentrates of reclaimed water. Because of these advantages, some of the assumptions on low-dose extrapolation are now being tested in fish.

In the past two decades, a significant body of research has developed on the relationship of chemical effects in fish compared to other experimental animals. Several fish species have been examined, medaka and trout most extensively (see, for example, Bailey et al., 1996; Bunton, 1996). The induction of cancer has received particular attention. Other recent research has focused on the similarity of the responses of aquatic species and mammalian species to endocrine disruption (Kavlock et al., 1996; Nimrod and Benson, 1996; Sumpter, 1995; Toppari et al., 1996).

Although fish have important advantages as toxicological test organisms, they also have disadvantages that are important to recognize. Disadvantages include the following:

1. Potentially important differences exist in the pharmacokinetics and metabolism of chemicals in fish as compared to mammalian species.

2. Acute responses of fish to chemicals in the water are unlikely to be representative of their effects on humans, reflecting the fact that a high fraction of such responses involves toxicity to the gill.

3. Certain mammalian functions are absent in fish, and certain functions in fish are not found in mammalian species.

4. Similar control mechanisms in both mammalian and fish species may control different physiological functions or control the same physiological function differently.

Some of these problems can be at least partly addressed by paying careful attention to the factors underlying various functions in mammalian and fish species. Measures of estrogenic responses in mammalian and fish species illustrate this relationship, as shown in Figures 5-1 and 5-2 (from Nimrod and Benson, 1996). As these figures illustrate, one significant difference between estrogenic responses in mammals (Figure 5-1) and fish (Figure 5-2) is that in fish, the liver plays a central role in transmitting signals to the ovary, whereas the liver plays no such role in mammals. In fish the liver secretes a protein, vitellogenin, that is ultimately involved in development of the egg yolk (Sumpter and Jobling, 1995). This difference between fish and mammals is derived from different responses to the hormone estradiol. In both fish and mammals, estradiol is secreted largely by the ovaries. In mammals, however, this hormone is responsible for expression of secondary sex characteristics, whereas in fish it regulates the secretion of vitellogenin from the liver.

Biochemically, then, the control of vitellogenin secretion in fish is analogous to those processes that determine sexual characteristics in mammals—two different outcomes stemming from the same essential

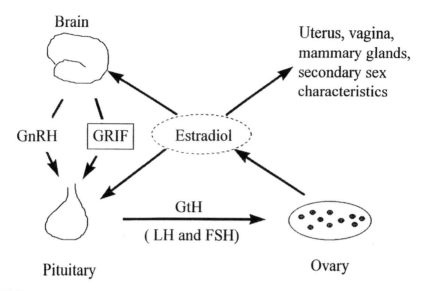

FIGURE 5-1 Estrogenic response in mammals: the hypothalamic-pituitary-gonadal axis. NOTE: GnRH = gonadotropin-releasing hormone; GRIF = gonadotropin inhibitory factor; GtH = gonadotropins.

biochemical dynamic. There are clear differences in the resulting physiology, but the basic underlying control mechanisms are the same. As a consequence, a chemical that alters vitellogenin secretion in fish is likely to have effects on the sexual characteristics of humans who consume the chemical at an effective dose.

Similar biochemical relationships between fish and mammals may affect the development of some health problems. For instance, carcinogens in general induce tumors in a variety of organs in mammalian species, while in fish the liver appears to be the primary target (Bailey et al., 1996; Bunton, 1996). This relationship could be significant for chemicals that act by tumor promotion and/or selective toxicity, as opposed to carcinogens that act by genotoxic mechanisms. Tumor promotion in particular tends to involve modifications in cell signaling systems responsible for the attraction of certain compounds or microorganisms to specific tissues or organs. Understanding the relationships of these signaling systems in fish and mammals would help answer whether liver tumor induction in fish would be predictive of tumor promotion in the development of breast cancer, for example. The parallels in endocrine disruption between fish and mammals suggest that this approach may not be as far-fetched as it first seems. Toxicity test systems using fish simply place

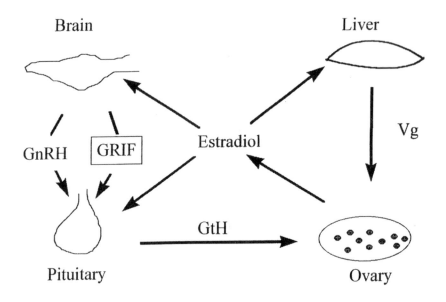

FIGURE 5-2 Estrogen response in fish: the hypothalamic-pituitary-gonadal-hepatic axis. NOTE: GnRH = gonadotropin-releasing hormone; GRIF = gonadotropin inhibitory factor; GtH = gonadotropins; Vg = vitellogenin.

more emphasis on understanding the mechanisms by which chemicals induce cancer.

The use of fish in testing the safety of reclaimed water would first require the following objectives to be met:

1. The relationship of known responses in fish and mammals would have to be better established. As an initial step, the existing database on toxicological tests in fish could be rigorously compared to toxicological results that have been obtained in rodent species and, where data exist, in humans. A better understanding of the basis of toxicological responses in fish and humans would also need to be established—at least with respect to mode of action and, where practical, to mechanism of action. Unfortunately, these data cannot be developed by testing of complex mixtures. Efforts should be aimed at individual chemicals that have toxicological properties and are likely to be encountered in the water column. Similar efforts are currently under way to develop an experimental base for linking human and rodent responses in more quantitative ways.

2. Likely routes of exposure would have to be considered and accounted for in interpretations of findings. For example, fish are sensitive

to highly lipophilic compounds of known toxicological importance (e.g., endocrine disrupters). However, humans are unlikely to be exposed to these chemicals via drinking water.

3. A research effort would have to be mounted to explore basic differences between the delivery of hydrophilic organic compounds and the delivery of hydrophobic organic compounds to the target organs in fish. So far, research in aquatic species has focused on hydrophobic compounds, which are not generally of importance in drinking water.

4. Scaling of the response variables in fish and mammalian species would have to be systematically undertaken. This is best done by examining very specific responses in one species with careful consideration of metabolic and pharmacokinetic differences. For example, the systemic doses of an endocrine disrupter needed to produce perturbations in estradiol levels could be examined in both fish and mammalian species. A follow-up could determine the levels of estradiol perturbation necessary to induce a change in uterine weight in rodents and an increase in vitellogenin secretion in fish.

Operational Considerations When Using Fish

Some critical practical considerations apply when using fish as the basic bioassay system. The interpretation of positive toxicological responses in fish must be confined to chronic rather than acute effects. The considerable body of data on acute toxic responses in fish involves relatively nonspecific toxic or irritant effects on the gill. In humans, the gastrointestinal tract greatly diminishes the importance of these irritant effects when exposure is through drinking water. Therefore, the wastewater must be of a quality that sustains fish for their normal life spans so that chronic effects can be examined.

The chronic effects monitored in fish should be selected based on their potential contribution to the development of disease in humans. This means that effects should be limited to (1) the development of a particular pathology (e.g., cancer, liver damage); (2) interference with specific physiological functions and processes (e.g., reproduction, development); or (3) molecular or biochemical effects that are recognized outcomes in both species. Without a clear connection to a recognized health effect, positive test results would be difficult to explain to the public. Once a valid end point is accepted, it should be understood that a positive response does, in fact, represent a hazard to human health unless new data indicate otherwise.

In some cases, secondary evaluations might show that a positive response in fish does not indicate a legitimate health risk for humans. For instance, it might be shown that the chemical causing the response in the test system does not reach the target site(s) in mammalian species at sufficient concentrations and is thus unlikely to do so in humans. Similarly, it might be demonstrated that the mechanism underlying the response is not found in humans, or that the intrinsic sensitivity of humans to the contaminant differs significantly from that of fish even after correcting for differences in internal dosimetry. However, such evaluations should be undertaken with the understanding that the assessment of health effects of unidentified or uncharacterized chemicals carries some inherent uncertainty.

Baseline testing of wastewaters with fish would need to be organized on a schedule that accounts for variations in water quality. Detection of day-to-day variability is both impractical and unnecessary for health threats from chronic effects. However, the assessment of seasonal changes in water quality may be useful. In-line tanks, either in series or in parallel, could be set up and fish harvested on a quarterly basis for examination of pathology and other indicators of adverse response. Studies that require less than three months to perform (e.g., developmental and reproduction experiments) could use separate tanks. In the case of chronic studies of carcinogenesis, it would be desirable to establish enough animals to allow sacrifices to be made on a quarterly basis. In order to obtain appropriate matching to the sampling requirements, a new group of fish would have to be introduced into the system each quarter.

Establishing relatively frequent sacrifices in groups with overlapping schedules provides important protection against marginal and/or spurious results. A positive result in one group can be validated to some extent by examining results with the group preceding or following it in time. If these results are consistent, the level of concern would be raised. If not, there would be sufficient justification for not addressing changes in treatment until the next quarter's results either confirmed or denied the positive result.

It is beyond the scope of this report to specify in detail the end points these studies should address. Obviously, routine observations should be made related to reproduction, development, and carcinogenesis. Other end points could be added, including immunotoxicity and neurotoxicity. However, it would be prudent to add such areas of concern in a systematic way. The complexity of the functions to be tested to assess various end points in fish and mammals will require careful consideration and development before the results can be properly interpreted.

Certain shorter-term tests will likely prove useful in the testing of treated wastewaters for particular end points of concern. However, these systems should not be considered as screening tools, but as tools to help identify chemicals that may be responsible for particular effects observed in the whole animal and then to trace the source of contamination. They may also be used to investigate physiological mechanisms of action more thoroughly.

The major advantage of such short-term systems is that their quicker results will allow pursuit of a problem on a timely basis. Municipal wastewater has multiple sources, and remedial action would require knowing which of these sources is potentially responsible for an observed health effect. A disadvantage of short-term tests for these purposes is that there is little assurance of congruence with the whole-animal test system (Arnold et al., 1996b), especially when unknown mixtures are being tested. Many unexplained interactions may have nonspecific origins for which there has been no scientific accounting. In particular, there are reports of synergistic responses in short-term systems that have no known connection to effects that have been observed in intact animals and/or human subjects (Arnold et al., 1996a).

In summary, there should be a clear decision path for toxicological testing of reclaimed water, using a test system that tracks live animals for a significant period of their life span. One possible approach is to use fish as the basis of testing, a topic that has been the focus of increasing research. If an effect is observed in the whole animal (such as the fish), risk would be estimated using general knowledge about the relative sensitivity of the animal and human systems. While these relationships will contain some uncertainties at the outset of testing, it should be possible to obtain refined estimates of the relative human susceptibility, if the test parameters are carefully thought out beforehand. More specific research can then be initiated to improve the risk assessment. Notwithstanding the difficulties of testing unknown chemical mixtures in reclaimed water, this decision path is quite viable in certain types of health outcomes or end points if the underlying basis of the response is understood (e.g., endocrine disruption). For some health outcomes, such as carcinogenesis, the mechanism is less well-understood, and it is probable that an observed effect will have to be accepted as implying an impact on human health. The alternative is to identify the specific chemical responsible for the observed effect and to reduce the risk associated with that chemical.

EPIDEMIOLOGIC STUDIES

In two locations, Los Angeles County and Windhoek, South Africa, potable reuse systems operational for some time have been the subjects of epidemiological studies. In addition, San Diego conducted baseline health statistics and an epidemiological feasibility study to assess the current population's health status and evaluate methods in anticipation of implementation of a proposed potable reuse project there. These studies are reviewed below.

Windhoek

Reclaimed water was first introduced into the Windhoek water system in November 1968 and was used sporadically after that. Augmentation of the water supply with reclaimed water was only practiced when drought conditions made it necessary. When reclaimed water was supplied, it was mixed with water from conventional sources (2 surface water impoundments and 36 boreholes scattered throughout the city prior to distribution). Therefore residents in areas that received reclaimed water had only periodic exposure to reclaimed water.

Study Design

Using an ecologic study design, epidemiologic studies of the health effects associated with the Windhoek reclaimed water supply began in 1973 and were expanded in 1976. Morbidity data were collected from private health care providers, hospitals, clinics, and health authorities. Diarrheal disease surveillance included the collection of data from every patient at a hospital, clinic, or general practitioner's office who had a Windhoek address. Mortality data were collected from death certificates in order to examine possible long-term health effects from the consumption of reclaimed water. When reclaimed water was not used in the system, baseline data on morbidity and mortality were collected. Disease rates for each ethnic group were compared between populations that lived in areas receiving reclaimed water and populations in areas with conventional water supplies.

Study Population

During the study period, the Windhoek population was 75,000 to 100,000 and was approximately 44 percent white, 44 percent black, and 12 percent "colored." The population was residentially segregated according to ethnic groups, and sanitation and hygiene conditions in the black and colored populations were considerably lower than in the white populations.

Exposure Assessment

Reclaimed water was supplied to some residential areas of each ethnic group, and much of the business district was served by reclaimed water. Exposure status was determined by place of residence. Investigators paid particular attention to children, since children tended to have higher rates of diarrheal disease and generally attended school close to their home, meaning they were likely to be exposed to one water supply.

Health Outcome Measurements

Investigators collected morbidity data on a number of infectious diseases (*Salmonella, Shigella,* enteropathogenic *Escherichia coli,* cholera, enterovirus, schistosoma, viral hepatitis, meningitis, encephalitis, and nonbacterial gastroenteritis). A single laboratory analyzed cultures from diarrheal cases for *Salmonella, Shigella, E. coli,* and *Vibrio cholerae* for the whole geographic area. Mortality data on diarrheal disease, tuberculosis, measles, diabetes, nutrition deficiencies, meningitis, hypertensive disease, ischemic heart disease, congestive cardiac failure, bronchitis/emphysema/asthma, cancer, and all other causes were collected from death certificates. The results of the health studies from 1976 to 1983 are based on the investigation of 15,000 episodes of diarrheal disease, 1000 cases of jaundice, and 3000 deaths.

Study Findings and Interpretation of Results

Only the most severe cases of diarrheal disease (those seeking medical care) were reported in this study. The majority (86 percent) of the diarrhea episodes reported at Windhoek occurred in children under 14 years of age. Diarrheal disease incidence was much higher in black and colored children than in white children, regardless of water supply. The major causes of mortality for whites were cardiovascular disease and cancer, whereas for blacks and coloreds, the leading causes were diarrheal disease and tuberculosis.

Epidemiolgists conducting the study statistically compared diarrheal disease incidence rates by year for whites living in areas receiving conventional water and those in areas receiving reclaimed water. Overall, they observed no significant difference over six years of observation. However, during two years (1977 and 1982), diarrhea incidence among whites (all ages) living in areas receiving water from conventional supplies was significantly higher than the incidence in whites living in areas receiving reclaimed water.

The investigators concluded that differences in diarrheal disease prevalence were "entirely related to socioeconomic factors and not to the nature of the water supply." As an example, they cited the morbidity patterns in two "socioeconomically similar groups of Windhoek residents." The group that was exposed to reclaimed water consistently had equal or lower incidence of diarrheal disease than the group that consumed water from conventional sources. Hepatitis A prevalence was also found to be related to general environmental conditions and personal hygiene rather than to water supply. The investigators did not attempt to relate mortality to water supply and felt that it would not be possible to do this in the future unless a "sudden unrecognized defect" in the water supply caused a "marked alteration in the mortality pattern" (Isaacson et al., 1987).

Strengths and Limitations

While the study attempted to compare possible health effects due to water across all ethnic and socioeconomic conditions, the small total population of Windhoek made it difficult to collect enough data (especially mortality data) to make statistically significant comparisons.

The results of this study are not generally applicable to populations in industrialized countries because of Windhoek's unique environment and demographic composition. The rate of diarrhea episodes reported in the Windhoek study from 1977 to 1982 ranged from 6 to 14 episodes per 1000 people per year in the white population drinking conventional water. In contrast, the rates of enteric infectious diseases reported to the Los Angeles County Health Department over the period 1989-1990 were 0.75 case per 1000 people per year in the population in the control areas and 0.9 case per 1000 people per year in the population using reclaimed water.

Orange and Los Angeles Counties Health-Effect Study

Two epidemiologic studies have been conducted by Orange and Los Angeles Counties in the Montebello Forebay area of Los Angeles County to examine health risks associated with exposure to reclaimed water used to replenish the area's ground water supply. The first study examined health outcomes from 1969 to 1980 (Frerichs, 1984; Frerichs et al., 1982), and the second one studied health outcomes from 1987 to 1991 (Sloss et al., 1986). Los Angeles County started recharging ground water supplies with treated wastewater in 1962, and for most systems in the study area, the percentage of reclaimed water increased over the following 30 years.

Study Population

The study population in both studies was estimated from census tract information for water districts and divided into areas known to use reclaimed water (exposed areas) and control areas, which were known to use water supplies that did not receive any reclaimed water but possessed demographic characteristics similar to the reclaimed water areas. In the first study, the population of the exposed areas of the Montebello Forebay region was between 478,000 and 486,000, according to the 1970 and 1980 censuses. The population in the control areas was between 677,000 and 576,000.

In the second study, the size of the population exposed to reclaimed water had almost doubled since the time of the first study, to around 900,000 people in the 1990 census. This represents about 10 percent of the total population of Los Angeles County. The control areas were in three parts of the county (parts of Montebello Forebay, Pomona, and northeastern San Fernando Valley) and included about 700,000 people.

Study Design

Both studies used an ecologic study design in which the unit of analysis was the census tract. Each census tract was assigned a categorical exposure variable derived from estimates of the percentage of reclaimed water in the water supply, as determined by the water suppliers serving the area. Information on health outcomes came from existing morbidity and mortality data collected by state and county surveillance programs, death certificates, cancer registries, and the University of Southern California Cancer Surveillance Program.

The first study also included a household survey in 1981 of randomly selected women in the two study areas. Telephone interviews were conducted with approximately 1200 adult females in "high" reclaimed water areas and 1300 women in the control areas to investigate possible differences in spontaneous abortions and other adverse reproductive outcomes, as well as other measures of general health (bed days, disability days, perception of well-being) and specific diseases. Data were collected on demographic and socioeconomic characteristics and possible confounding factors such as smoking and alcohol consumption, level of education, consumption of bottled water, and length of residence in the study area.

Exposure Assessment

The proportion of the water supply that originated from reclaimed water varied over time and geographic location. Each water supplier in the study area (32 in the first study, 39 in the second study) was queried about its water sources, delivery practices, service areas, and production levels during the study periods.

The actual amount of reclaimed water in the water supplies is unknown. To estimate the percentage of reclaimed water from each supplier, both studies used a model based on measurements of sulfate ion concentration in the ground water. Colorado River water, which has been used since 1954 to replenish the Montebello Forebay ground water basin, has high sulfate concentrations that allow its movement to be traced. Reclaimed water, by contrast, has no such characteristic mineral composition. However, the model assumes that the reclaimed water will follow a movement pattern similar to that of the Colorado River water in the ground water basin. In addition, the second study used regression analysis, the Kriging analytic method (which assumes that percentage of reclaimed water will be similar for neighboring wells), and analysis of travel-time contours to estimate the percentage of reclaimed water used between 1983 and 1990.

In the first study, estimates of overall reclaimed water concentrations ranged from 0 to 23 percent annually and from 0 to 11 percent over a 15-year period. Each census tract was categorized into one of four exposure categories (first study) or five exposure categories (second study) based on the estimated percentage of reclaimed water in the residential water supplies. Four broad exposure categories were used:

1. high reclaimed water areas, meaning areas that had received water containing 5 to 19 percent reclaimed water since 1969 or earlier;

2. low reclaimed water areas, meaning areas that had received water containing 0 to 4 percent reclaimed water before 1969;

3. central control areas, which had not received reclaimed water between 1962 and 1981; and

4. northwest control area, a nearby area outside of the study area but with similar demographic characteristics.

By the time of the second study, recharge with reclaimed water had been practiced for over 30 years, and estimates of the amount of reclaimed water used by many water systems had increased substantially. The maximum percentage of reclaimed water in 1990 was 31 percent, up from 19 percent in 1976. The annual maximum percentage of reclaimed water over the 30-year period varied considerably and ranged from less than 4 percent to between 20 and 31 percent. Data were obtained from 27 of the 39 water systems in the Montebello Forebay area. Twelve small water systems were excluded because their data either were unavailable or were considered unreliable for estimating the percentage of reclaimed water in the supplies. Five exposure categories were used, based on a 30-year average (1960-1990) of reclaimed water percentages:

1. RW1 (41 census tracts): average percentage of reclaimed water = 0.3
2. RW2 (28 census tracts): average percentage of reclaimed water = 1.5
3. RW3 (30 census tracts): average percentage of reclaimed water = 3.0
4. RW4 (23 census tracts): average percentage of reclaimed water = 6.5
5. RW5 (19 census tracts): average percentage of reclaimed water = 11.4

For the analyses using infectious disease health outcomes, the five exposure categories were based on three-year averages of the percentage of reclaimed water in order to reflect the shorter incubation period of infectious diseases. Three control areas—areas of 15, 21, and 81 census tracts in Los Angeles County—were chosen from water systems that had never used reclaimed water.

Health Outcome Measurements

In the first study, 21 health outcome measurements were used, based on data from three time periods: 1969 to 1971, 1972 to 1978, and 1979 to 1980. Measurements included the following:

- Mortality: all deaths, deaths from heart disease, from stroke, from all cancer, and from specific cancers (stomach, colon, bladder, rectum)
- Birth outcomes: low birth weight, infant and neonatal mortality, and congenital malformations
- Morbidity: incidence of stomach, colon, bladder, and rectum cancer; "potential waterborne diseases"; hepatitis A; and shigellosis

In the second study, 28 health outcome measurements were used, based on data from 1987 to 1991. Measurements included the following:

- Mortality data (based on death certificates from 1989-1991): all deaths, deaths from all cancer, from eight specific cancers (stomach, colon, bladder, rectum, esophagus, pancreas, liver, kidney), from heart disease, from cerebrovascular disease, and from all other causes
- Morbidity data (based on 1987-1991 cancer registry records): incidence of all cancers, of eight specific cancers (stomach, colon, bladder, rectum, esophagus, pancreas, liver, kidney), and infectious disease morbidity (based on reports to the Los Angeles County Health Department from 1989 to 1990, including incidence of *Giardia*, hepatitis A, *Salmonella*,

Shigella, amebiasis, cholera, gastroenteritis, leptospirosis, meningitis, and typhoid fever

Analytical Techniques

Both studies compared rates of these health outcomes, standardized by age and sex, to Los Angeles County rates. In the first study, differences in mortality and morbidity rates among the three exposure areas and one control area were tested by analysis of variance and weighted linear regression, followed by analysis of covariance or the Mantel-Haenszel nominal test of association. Depending on the analysis and availability of data, attempts were made to control for several confounding variables (age, sex, race, age of mother, and birth weight).

The second study used Poisson regression to calculate a rate ratio (morbidity or mortality rate in a reclaimed water area divided by the morbidity or mortality rate in the control area) and a confidence interval for each health outcome. In addition, sensitivity analyses were used to compare three regression models, controlling for (1) age and sex; (2) age, sex, and ethnicity; and (3) age, sex, ethnicity, and family income. All models controlled for population size.

Study Findings and Interpretation of Results

Overall, neither study observed consistently higher rates of either general or specific mortality or morbidity in the populations who lived in areas receiving higher percentages of reclaimed water.

The first study reported higher but statistically insignificant rectum cancer mortality rates in the three areas receiving reclaimed water compared to the control areas, and the area with the higher percentage of reclaimed water (5 to 19 percent) had greater rectum cancer mortality than the area with the lower percentage of reclaimed water (less than 5 percent). However, the number of rectum cancer deaths in these areas was relatively small. The investigators felt that the elevated levels of rectum cancer mortality would have been more meaningful if similar elevations in the mortality rates of stomach and colon cancer had also been observed. They attributed the elevated rectum cancer mortality rates to different death certificate coding practices by physicians in particular study areas (Sloss et al., 1996).

When examining the incidence rates of all potential waterborne infectious diseases, infectious hepatitis, and shigellosis, statistically significant differences were observed among the four study areas. However, the illness rates were highest in the control areas and lowest in the study area that received the highest percentage of reclaimed water. Also, significantly higher congenital malformations were observed in one of the control areas. The investigators did not comment on these findings.

In the second study, no consistent dose-response relationship was seen between exposure to reclaimed water and illness rates. The incidence of liver cancer was significantly higher (rate ratio = 1.7, 95 percent confidence interval = 1.1-2.7) in the study area that received the highest percentage of reclaimed water. However, no consistent dose-response relationship was observed in liver cancer incidence in the four other study areas that received lower percentages of reclaimed water. The incidence of stomach cancer and all cancers in the two study areas with the least reclaimed water was slightly higher than the rates of those illnesses in the control areas. Mortality from all cancers and from several specific cancers (liver, rectum, stomach, and kidney) was higher in reclaimed water areas compared to the control areas (rate ratios ranging from 1.12 to 1.71). However, most of these rate ratios were not statistically significant, did not show a consistent dose-response relationship, and showed very small magnitudes of differences between the experimental and control groups. The incidence of giardiasis, hepatitis A, and shigellosis was significantly higher in two study areas receiving low to medium percentages of reclaimed water. The infectious disease incidence rates in the study areas with the highest percentages of reclaimed water were either lower than the control areas or only slightly elevated.

In both studies, some significantly higher disease and/or mortality rates were observed both in some of the study areas that received reclaimed water and in some of the control areas. In their final interpretation of these results, the investigators considered the overall pattern of the results and whether the association between an exposure and a particular health outcome indicated a causal relationship, using commonly recognized criteria for causality (strength of the relationship, consistency, temporality, biologic gradient or dose-response, plausibility, and coherence). The absence of consistent dose-response relationships for all the elevated health outcomes that were observed in the study was the major argument against a causal association between exposure to reclaimed water and adverse health effects. The investigators concluded that the statistically significant associations that were observed occurred either due to chance, because of the large number of rate ratios that were calculated in the analyses, or due to differences between the exposed and control populations unrelated to the use of reclaimed water. The ecologic study design does not allow investigation of these differences.

Examination of the demographic characteristics of the study populations in the second study (from 1990 census information) indicates some differences among groups in ethnicity, education levels, and percentage employed in white-collar occupations. These demographic differences may affect risk factors for both chronic and acute diseases. The data analyses controlled for age, sex, and ethnicity differences between the reclaimed water areas and the control areas.

Strengths and Limitations

The major strengths of the OLAC studies are that they had large study populations and examined a large number or health outcomes. These are important features because one would expect the risks associated with reclaimed water exposure, if any, to be low, requiring a large study population to detect them. Also, the possible health risks associated with reclaimed water are undefined and might include both acute and chronic effects from microbial and chemical contaminants. Therefore it is advantageous to examine a broad array of health effects in a single study. The only health outcomes that were not included in these studies were reproductive outcomes (spontaneous abortions, birth defects, and infant mortality), which could be influenced by both acute and chronic exposures to contaminants in drinking water. One additional strength of the second study was that it could examine health effects that may be associated with long-term exposure or historical exposure to reclaimed water, since by the time the study was completed, reclaimed water had been used for about 30 years.

Because of the nature of the ecologic study design, both studies were unable to control for personal characteristics that might affect disease rates, such as smoking, diet, alcohol consumption, and occupational exposure. Ecologic studies assume that the study groups are roughly equivalent in all possible risk factors (except for the exposure of interest) for the health outcomes in question. These studies also assume that the exposure of interest will have the same effect on all the study groups (that is, that there will be no effect modification by group) (Sloss et al., 1996).

The investigators note that while the quality of the cancer incidence data used in these studies was high (due to a high-quality cancer registry), the quality of the mortality and infectious disease rate data was less than ideal (Sloss et al., 1996). Death certificates may not accurately record the cause of death or place of residence at time of death. The infectious disease surveillance by the Los Angeles County Department of Health Services may not report all diseases with the same accuracy. However, the quality of the health outcome data should be similar for both the exposed and the control populations.

Despite the large study populations in these studies, both the number of cancer illnesses over four years and the number of deaths from specific cancers over two years were very low compared to national averages. This is probably due to the relatively short study periods for these kinds of chronic diseases. Rectal cancer rates were particularly low, with only 36 deaths in the exposed populations and 21 deaths in the control areas from 1989 to 1991. For bladder cancer, there were only 48 deaths in the reclaimed water study areas and 35 deaths in the control areas. Therefore, in this study the risks associated with exposure to reclaimed water cannot be adequately evaluated for several important health outcomes because of the small number of events. Although the overall analyses do not suggest an association between adverse outcomes and reclaimed water in the drinking water supply, the public health significance of no or low rates of incidence, especially for specific outcomes, should always be interpreted in light of the statistical power of a study to detect an elevated risk. This caveat was not mentioned by the investigators.

The OLAC studies used a state-of-the-art model based on hydrogeologic and statistical theory to estimate the proportion of reclaimed water in the ground water at various times and locations. Attempts were made to compare estimates derived from the model with levels of constituents actually measured in the water. This model makes several assumptions and may have introduced an unknown amount of measurement error into the exposure estimates. In addition, the broad exposure categories used in these studies may not have been sufficiently different to show a clear dose-response relationship even if one was present. The studies were also unable to estimate actual exposure to reclaimed water based on personal differences in time spent away from home, consumption of bottled water and other beverages, and time lived in study area. The 1980 household survey in the first study indicated that 28 percent of respondents reported buying bottled water and 23 percent reported not drinking tap water (Frerichs, 1984). Given nationwide trends toward increased bottled water consumption during the 1980s, it is likely that the percentage of households using bottled water was higher in the second study.

Areas with highly mobile populations present difficulties in assessing long-term health risks. Data on population mobility in the "high reclaimed water" areas from the first study indicated that 40 percent of the population had lived in the same house for less than five years, and this was the most stable population of the four study areas (Frerichs et al., 1982). In the second study, the percentage of persons who had lived in the same house for less than five years ranged from 41 to 53 percent. Although the role of environmental exposures in the development of diseases with long latency periods (such as cancer) is a complex issue, it seems likely that 50 percent or more of the "exposed" study population would not have been exposed to reclaimed water long enough for reclaimed water to have an effect on cancer morbidity and mortality, since the minimum latency period for many cancers is believed to be about 15 years. The investigators acknowledged that exposure misclassification of people who recently moved into the study area would weaken the estimates of the effect of exposure on diseases with long latency periods (Sloss et al., 1996). Further, out-migration of "exposed" persons who had lived in the reclaimed water areas for long periods of time would reduce the statistical power of the study.

San Diego Feasibility Study

A pilot feasibility epidemiology study was conducted as a component of the San Diego Total Resource Recovery Project. The study sought to provide baseline health information on the residents of San Diego County that could be compared to future epidemiologic monitoring of health effects if potable reuse was adopted. This study also evaluated the feasibility, logistics, and cost of collecting various health data. Reproductive health and vital statistics (mortality and selected morbidity) were chosen by the investigators as "biological conditions that offer the most potential as environmental warning systems for environmental contamination" (Western Consortium for Public Health, 1996).

Study Design

The first element of this study was a survey of reproductive outcomes of women aged 15 to 44 in San Diego County. Telephone interviews were used to screen 19,504 residents for women eligible to participate in the reproductive health survey. From these interviews, approximately 1100 women were interviewed in their homes to collect demographic and health information: age, area of residence, self-perceived health, employment, height and weight, smoking exposures, alcohol consumption, diseases reported, income, ethnicity, and marital status. In addition, data pertaining to pregnancies were collected: weight gained, prenatal care, nausea, exposure to medication, diseases reported, duration, and pregnancy outcome.

The second element involved the collection of existing vital and health data routinely reported to San Diego County and the State of California. This included broad and specific mortality data, information on birth outcomes, and incidence of "various potential waterborne diseases" from 1980 through 1989.

The third element was a survey of neural tube birth defects using the California Birth Cohort Perinatal Files data from 1978 through 1985. The purpose of this survey was to establish baseline birth defects prevalence information in California as a whole and in San Diego County.

Study Findings

This epidemiologic feasibility study was a useful tool to evaluate methods for conducting further epidemiologic research in this field. The investigators concluded that while the vital statistics element was the least expensive component, it did not provide the necessary "precision in terms of data quality to address the health effects of wastewater recycling" (Western Consortium for Public Health, 1996). Instead, they recommended continuation of the neural tube defects study as the most cost-effective and reasonable approach to compare the rate of an appropriate health outcome in San Diego with that in the rest of California.

CONCLUSIONS AND RECOMMENDATIONS

Toxicology Studies

All six of the planned potable reuse projects reviewed in this chapter attempted to analyze the toxicological properties of reclaimed water. In most studies the testing was limited to mutagenic activity in bacterial systems, usually including at least two strains of the bacteria in the Ames *Salmonella* test. Some of the studies also used *in vitro* systems derived from mammalian cells, usually in the form of transformation assays and short-term *in vivo* clastogenesis assays (i.e., sister chromatid exchanges and micronucleus assays). Two projects employed chronic studies in live mammal systems to assess chronic toxicity, carcinogenicity, reproductive effects, and the potential for such waters to produce birth defects. Overall, the intent of toxicological testing can be grouped into (1) chemical screening and identification studies, (2) surveys of mutagenic activity, and (3) integrated toxicological testing.

The application of bioassays for screening and identifying chemicals that exhibit mutagenic activity, a methodology well accepted in the scientific community, was used in most of the more extensive studies. However, further toxicological evaluations are necessary in order to demonstrate whether or not the chemicals so identified have health effects.

The interpretation of the results from surveys of mutagenic activity is subject to the same ambiguity identified with other studies that have depended primarily on *in vitro* test systems. While the studies appeared to show no differences between reclaimed water and the conventional water source, positive results were obtained in screening assays, and there was no attempt to collect data in a system that would allow a rigorous comparison of relative risks associated with these two water sources.

Used alone, *in vitro* mutagenicity testing produces results of unclear meaning, because mutagenic activity alone does not imply carcinogenic risk to humans. A more accurate determination of health risk requires test systems that can more directly measure a complete range of health hazards and can use procedures for defining dose-response relationships to estimate the risks associated with varying levels of exposure. Such information cannot be derived from *in vitro* testing alone.

Of the toxicological studies, only the Denver and Tampa studies addressed a broad range of toxicological concerns. Those studies suggested that no adverse health effects should be anticipated from the use of Denver's or Tampa's reclaimed water as a source of potable drinking water. However, these studies, drawn from two discrete points in time and conducted only at a pilot plant level of effort, provide a very limited database from which to extrapolate to other locations and times.

Because of the high cost and methodological problems inherent in the testing of concentrated samples on rats and mice and because of the difficulty in applying the logic of safety testing to reclaimed water, the strategy set forth by the 1982 National Research Council panel is potentially too costly to implement and will not resolve health-effect questions in a timely manner for an operational potable reuse system. Accordingly, the committee recommends the following:

• **A new, alternative approach, such as the fish system described in this chapter, should be developed and used to continuously test the toxicity of reclaimed water in potable reuse projects.** The testing system described here should be viewed as a starting point for an approach that needs to evolve as deficiencies become apparent or as concerns with chemical contaminants focus on new health end points. It should employ a baseline screening test using a whole-animal rather than *in vitro* approach. The baseline screening tests should be conducted using water samples at ambient concentrations in order to reduce the uncertainty and high costs of using concentrates. The higher exposure possible with fish and the increased statistical power gained from using larger numbers of subjects will tend to offset the losses in sensitivity from not using high doses derived from concentrates.

• **Research efforts should be undertaken to understand the qualitative and quantitative relationships among responses in whole-animal test species, such as fish, and adverse health effects in humans.** *In vitro* short-term testing should be confined to qualitative evaluations of particular toxicological effects found in the product water in order to identify potential sources of contaminants and to guide remedial actions in a more timely manner. Such studies will probably have to employ concentration techniques.

• **For the fish testing system or any other toxicological test system used for reclaimed water, a clear decision path should be followed in**

toxicological testing. Testing should be conducted on live animals for a significant period of their lifespan. If an effect is observed in the whole animal, risk should be estimated using general knowledge about the relative sensitivity of the animal and human systems. While these relationships will contain some uncertainties at the outset of testing, it should be possible to obtain refined estimates of the relative human susceptibility if the test parameters are carefully thought out beforehand. More specific research can then be initiated to improve the risk assessment. Notwithstanding the difficulties of testing unknown chemical mixtures in reclaimed water, this decision path is quite viable for investigating certain types of health outcomes or end points if the underlying basis of the response is understood (e.g., endocrine disruption). For some health outcomes, such as carcinogenesis, the mechanism is less well-understood, and it is probable that an observed effect will have to be accepted as implying an impact on human health. The alternative is to identify the specific chemical responsible for the observed effect and to reduce the risk associated with that chemical.

• **The requirements for toxicological testing of water derived from an alternative source should be inversely related to how well the chemical composition of the water has been characterized.** If very few chemicals or chemical groups of concern are present, and the chemical composition of the water is well understood, the need for toxicological characterization of the water is low and may be safely neglected altogether. Conversely, if a large fraction of potentially hazardous and toxicologically uncharacterized organic chemicals is present, then toxicological testing will provide an additional assurance of safety.

Epidemiology Studies

Numerous epidemiologic studies (ecological, case-control, cohort, and outbreak investigations) have examined the relationship between various microbial and chemical contaminants in drinking water and a wide range of acute and chronic health outcomes in populations exposed either to a specific contaminated water supply or to specific types of source waters and treatment processes. However, only three such studies apply to potable reuse of reclaimed water, and only one set of epidemiological studies (Los Angeles County) evaluating the health effects associated with the consumption of reclaimed water has been conducted in a setting that is useful for assessing possible health effects in other parts of the United States or other industrialized countries. These studies have used an ecologic approach, which is appropriate as an initial step when the health risks are unknown or poorly documented, but negative results from such studies do not necessarily prove the safety of reclaimed water for human consumption. These studies can only be considered as preliminary examinations of the risks of exposure to reclaimed water.

The committee recommends that alternative epidemiologic study designs and more sophisticated methods of exposure assessment and outcome measurement be undertaken at a national level to evaluate the potential health risks associated with reclaimed water. Ecologic studies should be conducted in a variety of water reuse situations (e.g., ground water, surface water) in areas with low population mobility. Case-control studies or retrospective cohort studies should be undertaken to provide information on health outcomes and exposure for an individual level while controlling for other important risk factors. Although cohort studies are the most difficult and expensive to perform, this is the only study design that can examine the temporal relationship between exposure to reclaimed water and the development of adverse health effects. Increasing interest in and need for potable water reuse may justify such efforts.

REFERENCES

Anders, F., M. Schartl, and A. Branekow. 1984. Xiphophorus as an in vivo model for studies on oncogenes. National Cancer Institute Monograph 65:97-109.

Arnold, S. F., D. M. Klotz, B. M. Collins, P. M. Vonier, L. J. Guillette, Jr., and J. A. McLachlan. 1996a. Synergistic activation of estrogen receptor with combinations of environmental chemicals. Science 272:1489-1492.

Arnold, S. F., M. K. Robinson, A. C. Notides, L. J. Guillette, Jr., and J. A. McLachlan. 1996b. A yeast estrogen screen for examining the relative exposure of cells to natural and xenoestrogens. Environ. Health Persp. 104:544-548.

Ashby, J., and R. W. Tennant. 1988. Chemical structure, *Salmonella* mutagenicity, and extent of carcinogenesis as indicators of genotoxic carcinogenicity among 222 chemicals tested in rodents by the U.S. NCI/NTP. Mutation Research 24:17-115.

Bailey, G. S., D. E. Williams, and J. D. Hendricks. 1996. Fish models for environmental carcinogenesis: the rainbow trout. Environ. Health Persp. 104:5-21.

Baird, R. B., J. Gute, C. Jacks, R. Jenkins, L. Neisess, B. Scheybeler, R. van Sluis, and W. Yanko. 1980. Health effects of water reuse: a combination of toxicological and chemical methods for assessment. Pp. 925-935 in Jolley, R. L., et al. (eds.) Water Chlorination: Environmental Impact and Health Effects. Vol. 3. Ann Arbor Sci., Ann Arbor, Mich.

Baird, R. B., J. P. Gute, C. A. Jacks, L. B. Neisess, J. R. Smyth, and A. S. Walker. 1987. Negative-ion chemical ionization mass spectrometry and Ames mutagenicity tests of granular activated carbon treated waste water. Pp. 641-658 in Suffet, I. H., and M. Maliayandi (eds.) Advances in Chemistry series 214. Organic Pollutants in Water: Sampling, Analysis and Toxicity Testing.

Bull, R. J., and F. D. Kopfler. 1981. Toxicological evaluation of risks associated with potable reuse of wastewater. Pp. 2176-2194 in Proceedings of Water Reuse Symposium II - Vol. 3, August 23-28, Washington, D.C. Denver, Colo.: American Water Works Association Research Foundation.

Bull, R. J., L. S. Birnbaum, K. P. Cantor, J. B. Rose, B. E. Butterworth, R. Pegram, and J. Tuomisto. 1995. Symposium overview: water chlorination: essential process or cancer hazard. Fund. Appl. Toxicol. 28:155-166.

Bunton, T. E. 1996. Experimental chemical carcinogenesis in fish. Toxicologic Pathology 24:603-618.

Calabrese, E. J., L. A. Baldwin, et al. 1992. Epigenetic carcinogens in fish. Review Aquatic Science 6(2):89-96.

Cheh, A. M., J. Skohdopole, P. Koski, and L. Cole. 1980. Nonvolatile mutagens in drinking water, production by chlorination and destruction by sulfite. Science 207:90-92.

CH_2M Hill. 1993. Tampa Water Resource Recovery Project: Pilot Studies. Tampa, Fla.: CH_2M Hill.

Courtney, L. A., and J. A. Couch. 1984. Usefulness of *Cyprinodon variegatus* and *Fundulus grandis* in carcinogenicity testing: advantages and special problems. National Cancer Institute Monograph 65:83-96.

Food and Drug Administration (FDA). 1982. Toxicological Principles for the Safety Assessment of Direct Food Additives and Color Additives Used in Food (Red Book). Washington, D.C.: Food and Drug Administration, Department of Health and Human Services.

Frerichs, R. R. 1984. Epidemiologic monitoring of possible health reactions of wastewater reuse. Science of the Total Environment 32:353-363.

Frerichs, R. R., E. M. Sloss, and K. P. Satin. 1982. Epidemiologic impact of water reuse in Los Angeles County. Environmental Research 29:109-122.

Hatanaka, J., N. Doke, et al. 1982. Usefulness and rapidity of screening for toxicity and carcinogenicity of chemicals in medaka, *Oryzias latipes*. Japan J. Exp. Med. 52(5):243-253.

Hawkins, W. E., R. M. Overstreet, et al. 1984. Tumor induction in several small species by classical carcinogens and related compounds. Fifth Conference on Water Chlorination: Environmental Impact and Health Effects. Williamsburg, Va.: Lewis Publishers, Inc.

Hawkins, W. E., R. M. Overstreet, et al. 1985. Development of aquarium fish models for environmental carcinogenesis: tumor induction in seven species. J. Appl. Toxicol. 5(4):261-264.

Hemmer, J., C. Hamann, and D. Pickard. 1994. Tampa Water Resource Recovery Project. San Diego Water Reuse Health Effects Study. 1994 Water Reuse Symposium Proceedings, Feb. 27-Mar. 2, Dallas, Tex. Denver, Colo.: American Water Works Association.

ILSI. 1995. Disinfection by-products in drinking water: Critical issues in health effects research. Pp. 110-120 in Workshop Report. Chapel Hill, N.C. Oct. 23-25.

Isaacson, M., and A. R. Sayed. 1988. Health aspects of the use of recycled water in Windhoek, SWA/Namibia, 1974-1983. South African Medical Journal 73:596-599.

Isaacson, M., A. R. Sayed, and W. H. J. Hattingh. 1987. Studies on health aspects of water reclamation during 1974-1983 in Windhoek, South West Africa/Namibia. Report No. 38/1/87. Pretoria: Water Research Commission.

Jacks, C. A., J. P. Gute, L. B. Neisess, R. J. Van Sluis, and R. B. Baird. 1983. Health effects of water reuse: characterization of mutagenic residues isolated from reclaimed, surface, and groundwater supplies. Pp. 1237-1248 in Jolley, R.L., et al. (eds.) Water Chlorination: Environmental Impact and Health Effects. Vol. 4. Ann Arbor, Mich.: Ann Arbor Sci.

James M. Montgomery, Inc. 1983. Operation, Maintenance, and Performance Evaluation of the Potomac Estuary Wastewater Treatment Plant. Alexandria, Va.: Montgomery Watson, Inc.

Jenkins, R. L., C. A. Jacks, R. B. Baird, B. J. Scheybeler, L. B. Neisess, J. P. Gute, R. J. Van Sluis, and W. A. Yanko. 1983. Mutagenicity and organic solute recovery from water with a high-volume resin concentrator. Water Res. 17:1569-1574.

Kavlock, R. J., G. P. Daston, C. DeRosa, et al. 1996. Research needs for risk assessment of health and environmental effects of endocrine disruption: a report of the U.S. Environmental Protection Agency sponsored workshop. Env. Health Persp. 104:715-740.

Kool, H. J., C. F. van Kreijl, E. deGreef, and H. J. van Kranen. 1982. Presence, introduction and removal of mutagenic activity during the preparation of drinking water in the Netherlands. Environ. Health Persp. 46:207-214.

Krause, M. K., L. D. Rhodes, et al. 1997. Cloning of the p53 tumor suppressor gene from the Japanese medaka (*Oryzias latipes*) and evaluation of mutational hotspots in MNNG-exposed fish. Gene 189(1):101-106.

Lauer, W. C. 1993. Denver's Direct Potable Reuse Demonstration Project Final Report—Executive Summary. Report prepared by the Denver Water Department, Denver, Colo.

Lauer, W. C., and S. E. Rogers. 1996. The demonstration of direct potable water reuse: Denver's pioneer project. Pp. 269-289 in AWWA/WEF 1996 Water Reuse Conference Proceedings, San Diego, Calif., February 25-28. Denver, Colo.: American Water Works Association.

Lopez, L., and A. DeAngelo. 1997. Carcinogenicity of dichloroacetic acid in the Japanese medaka small fish model. 18th Annual Meeting of the Society of Environmental Toxicology and Chemistry, San Francisco, Calif.

Meier, J. R., R. D. Lingg, and R. J. Bull. 1983. Formation of mutagens following chlorination of humic acids: a model for mutagen formation during drinking water treatment. Mutation Res. 18:25-41.

Meier, J. R., R. B. Knohl, W. E. Coleman, H. P. Ringhand, J. W. Munch, W. H. Kaylor, R. P. Streicher, and F. C. Kopfler. 1987. Studies on the potent bacterial mutagen, 3-chloro-4-(dichloromethyl)-5-hydroxy-2(5H)-furanone: aqueous stability, XAD recovery, and analytical determination in drinking water and in chlorinated humic acid solutions. Mutation Res. 189:363-373.

National Research Council (NRC). 1982. Quality Criteria for Water Reuse. Washington, D.C.: National Academy Press.

National Research Council (NRC). 1984. The Potomac Estuary Experimental Water Treatment Plant. Washington, D.C.: National Academy Press.

Nellor, M. H., R. B. Baird, and J. R. Smyth. 1984. Health Effects Study Final Report. County Sanitation Districts of Los Angeles County, Whittier, Calif.

Nimrod, A. C., and W. H. Benson. 1996. Environmental estrogenic effects of alkylphenol ethoxylates. Critical Rev. Toxicol. 26:335-364.

Pereira, M. A., B. C. Casto, M. W. Tabor, and M. D. Khoury. Undated. Toxicology Studies of the Tampa Water Resources Project. Report supplied to the committee by M. Pereira, Medical College of Ohio, Toledo.

Sato, A., J. Komura, et al. 1992. Firefly luciferase gene transmission and expression in transgenic medaka (*Oryzias latipes*). Mol. Mar. Biol. Biotechnol. 1(4-5):318-325.

Sloss, E. M., S. A. Geschwind, D. F. McCaffrey, and B. R. Ritz. 1996. Groundwater Recharge with Reclaimed Water: An Epidemiologic Assessment in Los Angeles County, 1987-1991. Santa Monica, Calif.: RAND.

Sumpter, J. P. 1995. Feminized responses in fish to environmental estrogens. Toxicology Lett. 82/83:737-742.

Sumpter, J. P., and S. Jobling. 1995. Vitellogenesis as a biomarker for estrogenic contamination of the aquatic environment. Environ. Health Persp. 103(Suppl. 7):173-178.

Toppari, J., J. C. Larsen, P. Christiansen, et al. 1996. Male reproductive health and environmental xenoestrogens. Environ. Health Persp. 104:741-803.

Tuomisto, J., J. Hyttinen, K. Jansson, H. Komulainen, V. L. Kosma, J. Maki-Paakkanen, S. L. Vaittinen, and T. Vartianen. 1995. Pp. 30-31 in Disinfection By-Products in Drinking Water: Critical Issues in Health Effects Research. Workshop Report. Chapel Hill, N.C. Oct. 23-25. Washington, D.C.: ILSI.

U.S. Environmental Protection Agency (EPA). 1979. Proposed health effects testing standards for Toxic Substances Control Act test rules: Environmental Protection Agency. Federal Register 44:44054.

Walker. W. W., C. S. Manning, et al. 1985. Development of aquarium fish models for environmental carcinogenesis: an intermitent-flow exposure system for volatile hydrophobic chemicals. Journal of Applied Toxicology 5(4):255-260.

Western Consortium for Public Health. 1992. The City of San Diego Total Resource Recovery Project: Health Effects Study—Final Summary Report. Oakland, Calif.: Western Consortium for Public Health.

Western Consortium for Public Health. 1996. Total Resource Recovery Project Final Report, City of San Diego. Western Consortium for Public Health in association with EOA, Inc. Oakland, Calif.: Western Consortium for Public Health.

6

Reliability and Quality Assurance Issues for Reuse Systems

In recent years, the term "barriers" has come to serve as a comprehensive descriptor of processes that tend to reduce the risk of waterborne contaminants. Watershed protection programs, water treatment processes, and maintenance of the water distribution infrastructure are all considered barriers to certain types of contamination. The concept of barriers is attractive because it promotes integrated thinking about actions that at first appear unrelated. Potable reuse projects require more robust multiple barriers than conventional water systems do, especially for microbiological contaminants. For example, the presence of high concentrations of *Cryptosporidium* in a water supply for even a very short period poses a significant risk, whereas high levels of lead would be expected to cause detrimental effects only if the lead persisted for much longer periods of exposure.

As an engineered system, every water reclamation facility has the potential for out-of-specification performance. This chapter considers the practical management alternatives for reducing the risks posed by the potential for such performance failures. The chapter also considers the role of public health surveillance in providing an early warning system for potential health effects.

MULTIPLE BARRIERS

From 1946 until 1980, 41.7 percent of drinking water disease outbreaks in community water supplies were attributed to inadequate or interrupted treatment (Lippy and Waltrip, 1984), and similar treatment failures have been noted in more recent years (Herwaldt, 1991). Multiple barriers to contaminant breakthrough are seen as one way of preventing such outbreaks. In addition, increased concern over resistant viruses and protozoa has led the water treatment community to pay more attention to the concept of incorporating multiple barriers to such pathogens in the water system.

Velz (1970) employed the term "multiple barriers" for the concept of providing wastewater treatment when a receiving water is used for water supply. In conventional (non-reuse applications) water treatment, the use of multiple barriers to pathogens within a single facility was advocated by the American Water Works Association Organisms in Water Committee (1987). The multiple barriers concept has, in effect, been embodied in the federal Surface Water Treatment Rule, as well as a number of the state-specific reuse requirements noted earlier in this report. Currently, all processes that help to reduce the risk of waterborne contaminants are referred to as barriers. Thus, watershed protection programs, engineered water treatment processes such as disinfection and filtration, and maintenance of the water distribution infrastructure are all considered barriers to certain types of contamination (even though some of these might cause certain aspects of water quality to deteriorate).

Different types of contamination require different types of barriers. For instance, disinfectants aimed at microbial pathogens do not mitigate chemically based risks and in fact may exacerbate them. Likewise, activated carbon will remove many chemical contaminants but does little to remove viruses. Accordingly, each barrier must be examined separately for its efficacy for removal of each contaminant. The cumulative capability of all barriers to accomplish removal should be evaluated considering the levels of the contaminants in the source water, the nature of the expected health effect associated with the contaminants, the goals that have been set for the potable supply, and any additional safety factors.

The concept of multiple barriers is implicit in the design of many advanced wastewater treatment projects investigating the feasibility of reuse. These projects typically use several physical and chemical barriers to pathogens, which can cause problems if present at high levels for even a short time, but only one or two aimed at chemical contaminants, which must generally be present longer to affect health. One such design is the San Diego project, where a failure in the ion exchange process would probably cause the finished water to exceed the nitrate target until the ion exchange unit was repaired, but where there is no one process whose failure would prevent virus goals from being met.

For drinking water supplies in general, using multiple barriers might mean choosing the most pristine available water source, protecting it from current and future contamination, providing multiple engineering processes to remove contaminants in a water treatment plant (e.g., pre-oxidation, coagulation, filtration, and disinfection), and protecting the water quality from deterioration in the distribution system (e.g., by adding corrosion inhibitors and additional disinfectant and keeping the water under pressure to prevent contaminated water from entering the system). In places where reclaimed water is used to augment natural supplies or where source water cannot be protected from upstream discharges of water with impaired quality, the importance of the barriers at the water treatment plant or wastewater reclamation plant is correspondingly increased.

EVALUATING BARRIER INDEPENDENCE

The independence of multiple barriers is a key aspect of system reliability and safety, especially for removal of pathogens. The greater the degree of independence among different barriers, the more they can be relied upon to serve as backups for one another. The water treatment process train should incorporate multiple, independent treatment barriers of sufficient redundancy that required contaminant removal levels will be achieved even if the single most effective treatment barrier is not performing. For example, sedimentation and filtration should not be considered independent barriers if the success or failure of both depends on proper coagulation prior to the sedimentation step. Failure in this type of system contributed to the 1993 *Cryptosporidium* outbreak in Milwaukee. On the other hand, design of a sufficiently deep intake pipe for surface water extraction and the use of disinfection are independent barriers to microbial contamination.

Individual treatment barriers (or unit processes) should be evaluated individually and collectively with regard to their capacity for contaminant removal and their prospects for failure. Analysis of the contaminant-removal capabilities of independent barriers involves several steps. First, the contaminants of concern are identified, and reasonable maximum and target levels are determined. (The difference between target level and maximum level provides a margin of error.) Based on information on contaminant levels in the source water, the necessary contaminant reduction, usually expressed as logs of removal, is estimated. Each treatment process is evaluated for its removal capability, and this information is used to estimate the overall removal to be accomplished by the process train. Finally, an estimate is made for how much the overall removal will be compromised if the single most important independent barrier (unit process) were to fail.

Table 6-1 illustrates this analysis for a hypothetical comparison of two treatment trains. The table compares overall efficiencies of the process trains for removing four hypothetical microbial pathogens (A, B, C, and D). This hypothetical analysis illustrates that both process trains provide similar protection against pathogens A and B, but the train that includes the membrane process provides considerably greater protection against pathogens C and D.

USE OF ENVIRONMENTAL BUFFERS IN REUSE SYSTEMS

By definition, indirect potable reuse projects include an "environmental buffer," that is, a natural water body that physically separates the product water from the wastewater reclamation plant and the intake to the drinking water treatment plant. A reservoir, river, or lake would be the environmental buffer for planned surface water augmentation. With ground water augmentation, the aquifer and/or soil (depending on whether direct injection is used) acts as the environmental buffer between the reclamation plant and the water production well. Surface water always receives subsequent treatment prior to distribution, while ground water may or may not.

The effectiveness of environmental buffers as barriers to various types of contamination is less well understood than that of engineered treatment processes. In different wastewater reuse applications the environmental barriers might include dispersion, dilution, sorption to the sediment and removal by deposition, chemical reaction, biodegradation, and biological transformation processes (photolysis and hydrolysis). Analyzing the effectiveness of environmental buffers for contaminant removal is complex, for a number of reasons. There are different removal processes for different contaminants, and different processes are expected to dominate depending on whether the water infiltrates through surface layers of the ground, is injected into deeper underground layers, or is discharged to a surface water reservoir. Even if all controllable factors are identical, local geology, biology, and climate undoubtedly affect the outcome.

In the absence of definitive research on the effectiveness of environmental buffers, attitudes about this topic have developed from a combination of anecdotal evidence and attempts to extrapolate from the behavior of other systems. The three perceived benefits of buffers are that they provide (1) an opportunity to further reduce contaminants through natural processes, (2) a substantial lag time between the exit of the advanced wastewater treatment system and entrance into the potable system, and (3) the opportunity for the water of wastewater origin to blend with natural waters in the environment. These three potential benefits are discussed below.

TABLE 6-1 Hypothetical Comparison of Two Multiple-Barrier Treatment Process Trains

	Pathogen A	Pathogen B	Pathogen C	Pathogen D
Physical/Chemical Process Train				
Secondary effluent	1.00E + 07	1.00E + 01	1.00E + 02	1.00E + 01
Nominal "safe" level	1.00E + 00	1.00E − 05	1.00E − 05	1.00E − 05
Safety factor	1.00E + 02	1.00E + 03	1.00E + 03	1.00E + 03
New goal	1.00E − 02	1.00E − 08	1.00E − 08	1.00E − 08
Required logs removal	9	9	10	9
Treatments				
Lime	1.7	2.0	0.0	0.0
Recarbonation	0.0	0.0	0.0	0.0
Filtration	0.3	0.3	0.3	0.3
GAC[a]	1.0	1.0	0.3	0.3
AOP[b]	5.0	6.0	3.0	1.5
UV[c]	4.0	3.0	2.0	1.0
Cl_2[d]	4.0	5.0	2.0	0.0
Total logs removal	16	17	8	3
Logs w/o key barrier	11	11	5	2
Excess logs	2	2	−5	−7

Process Train with
Membrane Filtration

Secondary effluent	1.00E + 07	1.00E + 01	1.00E + 02	1.00E + 01
Nominal "safe" level	1.00E + 00	1.00E − 05	1.00E − 05	1.00E − 05
Safety factor	1.00E + 02	1.00E + 03	1.00E + 03	1.00E + 03
New goal	1.00E − 02	1.00E − 08	1.00E − 08	1.00E − 08
Required logs removal	9	9	10	9
Treatments				
MF[e]	5.0	0.5	5.0	5.0
RO[f]	4.0	4.0	5.0	5.0
Stripper	0.0	0.0	0.0	0.0
AOP[b]	5.0	6.0	3.0	1.5
UV[c]	4.0	3.0	2.0	1.0
Cl_2[d]	4.0	5.0	2.0	0.0
Total logs removal	22	19	17	13
Logs w/o key barrier	17	13	12	8
Excess logs	8	4	2	−2

[a]GAC = granular activated carbon filtration.
[b]AOP = advanced oxidation process.
[c]UV = ultraviolet disinfection.
[d]Cl_2 = chlorination.
[e]MF = microfiltration.
[f]RO = reverse osmosis.

Reduction of Contaminants Through Natural Processes

The value of the soil-aquifer system in attenuating pathogens is long known. During the middle of the nineteenth century, scientists and doctors observed that epidemics of cholera and typhoid were more common in cities served by river water than in cities served by ground water. These observations stimulated the development of "natural filters," a concept similar to the filtration of drinking water through sand banks as practiced in Germany today (Sontheimer, 1980). These techniques reportedly worked well in controlling bacterial pathogens. Soil-aquifer systems have demonstrated substantial capacity to remove bacterial pathogens and organic chemicals (NRC, 1994). The ability of aquifer systems to reduce viruses is less certain, and in some instances, viruses have been reported to survive transport for long distances in ground water (NRC, 1994).

The differences between surface water and ground water storage as environmental buffers are profound, both in terms of processes that may reduce the levels of contaminants of concern and in terms of the degree to which the environmental buffer can be influenced by short circuiting. The potential for short-circuiting of flows introduced into a surface water body deserves special attention.

More often than not, surface water bodies are vertically stratified due to differences in density. Temperature is usually the primary driver of this density difference, although salinity also plays a role. Raphael (1962) demonstrated the importance of depth in determining mixing in reservoirs. He characterized the reservoir as a series of horizontal slabs, each corresponding to a different density layer. Since then it has been shown that this density stratification is extremely stable, that very little mixing occurs between stratified layers, and that a small temperature difference will induce stratification (Fischer et al., 1979). Many reservoirs are stratified in this way for most of the year. In warm climates, stratification may persist nearly year-round.

Figure 6-1(a) shows the structure of the Wellington Reservoir during the summer months. The Wellington Reservoir is modeled as four slabs of water covering a temperature range of about 3 to 4°C each. The top slab (23-26°C) is the epilimnion and extends from a depth of 0 to 15 m. The second slab (19-22°C) is the upper half of the thermocline, which is a much thinner layer at a depth between 15 and 16 m. The third slab (16-19°C) represents the bottom half of the thermocline, another thin layer located 16-18 m deep. The fourth and last slab (13-15°C) is the hypolimnion, located between 18 and 30 m of depth.

Water introduced into a reservoir will travel vertically up or down to the slab that corresponds to its temperature, after which it will spread horizontally. Horizontal mixing within the slabs is easily accomplished. Vertical mixing between slabs is very limited, hence the term "slab." In the reservoir model shown in Figure 6-1(b), an influent water at a temperature of 23°C would mix throughout slab 1 (epilimnion) and be diluted by mixing with nearly half the contents of the reservoir. An influent water at a temperature of 21°C, on the other hand, would mix only within the much smaller percentage of reservoir water in slab 2 (the upper thermocline).

The location and depth of reservoir inlets do not heavily influence this dynamic. No matter what depth water is introduced, it will seek water of its own density. Cold water discharged into the warm layers on the top will sink; warm water discharged into the cold layers on the bottom will rise.

In contrast, the depth of the reservoir outlet can have an important impact on whether the reclaimed water finds a "short-circuit" through the natural system, preventing it from receiving all the benefits of natural treatment. In Figure 6-1(b), if the reservoir's outlet is 15.5 m deep and water is discharged into the reservoir at 21°C, the mixing will be only within slab 1, and the discharged water will move almost directly to the outlet. The result will be a very serious short-circuiting problem. Tests conducted at Lake Youngs for the Seattle Water Department under such conditions showed that water marked by tracers passed through the system in 31 hours—even though the lake's nominal detention time was about 30 days (Ellis, 1995).

In the Wellington example, if the outlet is placed at a depth of 25 m in slab 4, short-circuiting can be prevented for a very long time because of the poor circulation between slabs. Tracer tests conducted under such conditions in the San Vicente Reservoir for the City of San Diego (Flow Science for Montgomery Watson and City of San Diego, 1995) showed that less than 0.2 percent of the tracer had passed through in 30 days. At the time, this reservoir had a hydraulic detention time of approximately 12 months. This strategy of discharging to the epilimnion (top layer) and withdrawing from the hypolimnion (bottom layer) is often a sound one for indirect potable reuse projects. One of the principal reasons for this is the fact that treated wastewater is usually as warm or warmer than the epilimnion for most of the year.

Thus, the holding time that can be accomplished in a surface water body can be much less than it might appear. While such a water body might appear to be well mixed, it is more likely stratified into numerous horizontal layers of different densities. Calculations of the hydraulic detention time may have little relevance because water discharged to the reservoir will seek a layer of equal buoyancy and mix with that layer

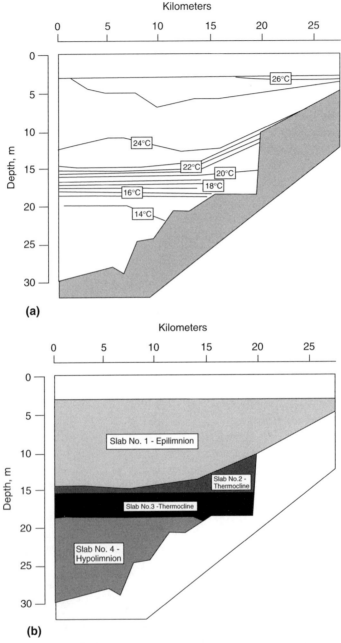

FIGURE 6-1 Wellington Reservoir. (a) Temperature profiles in midsummer. (b) Division into four slabs, each with a temperature range of 4°C. SOURCE: Reprinted, with permission, from Fischer et al., 1979. © 1979 by Academic Press.

only. If water mixes and is also withdrawn from the thin thermocline, short-circuiting can be surprisingly rapid. Therefore, water repurification projects using surface water bodies as environmental buffers should be careful to locate drinking water intakes so that any introduced wastewater will have to pass through several layers of water to reach the intakes. It should not be assumed that long hydraulic detention times will control short-circuiting. Investigations into the stratification of surface water bodies during the year are critical to designing effective ways to use them as environmental buffers.

Little study has been devoted to the removal of contaminants from water during transport or storage in surface water bodies. The limited literature on virus removal suggests some removal is possible. The discussion in Chapter 3 on microorganism survival in ambient water notes the importance of temperature. The literature also suggests that the removal of some particularly volatile synthetic organic contaminants can be expected during storage in a large surface water body, but the committee is unaware of data demonstrating removal of disinfection by-product precursors or other organic components of health concern.

In summary, present evidence suggests that some contaminant removal does occur during storage in or passage through environmental buffers for soil-aquifer treatment systems and in aquifer and surface water systems where residence times are long and short-circuiting is avoided. On the other hand, the performance of environmental buffers has not been extensively documented, the conditions that benefit or confound performance are poorly understood, and the necessary data on underground conditions are difficult to obtain. As a result, while environmental buffers can be expected to play a role in public health protection, particularly from microbial disease, the level of protection provided is difficult to quantify.

Lag Time Between Discharge of Wastewater and Entry Into the Potable Water System

When short-circuiting is controlled, using a large environmental buffer provides a significant lag time between the time the reclaimed water is produced and the time it enters the domestic supply. This increased lag time allows any processes contributing to contaminant reduction to proceed longer and, presumably, accomplish more removal.

However, the lag time should be viewed as simply one factor affecting the contaminant removal processes described above. A more subtle yet important value of lag time is that it allows flexibility in responding to changes in the quality of the treated water that are not understood at the time the water is produced but become clear later. For instance, a substantial lag time increases the chance that a flaw in treatment performance that goes undetected at one time will be discovered before the affected water enters the potable water system. There is always value in having additional time to react. This extra precaution might become less necessary after decades of experience with potable reuse systems, but at present it seems quite valuable.

On the other hand, some argue that storage time accomplishes nothing, that it only ensures that the project will cost more and that if contaminants are inadvertently released, they could affect water supplies over a longer term. Admittedly the alternatives for action are fewer when water of poor quality has been discharged into a supply lake or aquifer. Yet even when the quality of the reclaimed water is poor, mixing it with higher quality water that is subsequently treated is probably a better option than having that same poor quality enter the potable system more quickly with little or no dilution and time lag. As long as the water is in storage, measures can be taken to protect the public. In California, regulations have been proposed requiring 6 to 12 months of hydraulic retention time in an environmental buffer before treated wastewater can be reused as a potable water source. These storage times seem appropriate for allowing the development and execution of alternative actions if needed.

Loss of Identity of the Water of Wastewater Origin

"Loss of identity," meaning mixing of the reclaimed water with water in a natural system, is certainly an important element in the conceptual view of water reuse by both public health authorities and the average citizen. When reclaimed water is introduced into a large aquifer, lake, or further upstream in a river, the process of water reuse is less visible.

Nevertheless, loss of identity alone would seem to have little serious value where public health protection is concerned, because it is hard to associate the loss of identity with any tangible reduction in risk. In the absence of other documented conversion or removal processes, loss of identity is simply dilution. While there might be some risk reduction associated with such dilution, an honest appraisal of the benefit of loss of identity should be presented in those terms.

Moreover, the loss-of-identity argument is not always well received by the public. This can be particularly true when the quality of the reclaimed water is perceived to exceed the quality of the water presently in the environmental buffer. Under these conditions, blending the repurified water with these sources can be seen as paying a premium to satisfy arbitrary requirements while risking contamination of the high-quality reclaimed water. This was one of the reactions received from public workshops dealing with water reuse in San Diego, where there are plans to introduce high-quality reclaimed water into Lake San Vicente.

In summary, it is clear that including an environmental buffer in a potable reuse project can substantially reduce public health risk but that this risk reduction cannot presently be assessed with any confidence. The benefits that do accrue are likely to be associated with reduction in contaminant concentration and the introduction of a lag time. For these benefits to be realized, short-circuiting must be avoided. Aquifers appear to provide more protection as environmental buffers than do surface waters. Soil-aquifer treatment adds a further dimension of potential contaminant reduction. When surface water storage alone is used, more sophisticated treatment and longer storage times should be required. Finally, loss of identity is an issue that seems more relevant to public relations than public health protection. However, an environmental buffer large enough to contribute to a loss of identity is probably also large enough to bring real reductions in risk.

CONSEQUENCE-FREQUENCY ASSESSMENT

A water reclamation plant is an engineered system. Environmental engineering texts have stated that "every water treatment facility must be so designed that, when properly operated, it can produce continuously the design rate of flow and meet the established water quality standards" (James M. Montgomery Consulting Engineers Inc., 1985). However, any engineered system has the potential for out-of-specification performance. It is critical to understand the likelihood of failures that may compromise product safety. The tools and concepts used to analyze reliability of other engineered systems should also be used in analyzing potable reuse systems (Kumamoto and Henley, 1996).

Regardless of the treatment processes used, treated drinking water from a given facility will vary in quality because of variations in the influent stream as well as variability in the performance of the individual process elements. For contaminants (such as carcinogens) associated only with long-term health risks, variability is relatively unimportant, since the effect depends primarily upon the long-term average dose. However, variability becomes quite important for contaminants, such as infectious microbes or acute chemical toxins, that cause acute health effects at high or single-dose exposures. Figure 6-2 presents two hypothetical data sets showing concentrations of a hypothetical contaminant. In both cases the mean concentration is 3. If the contaminant were one that caused an acute effect when present above a concentration of 4, then the high-vari-

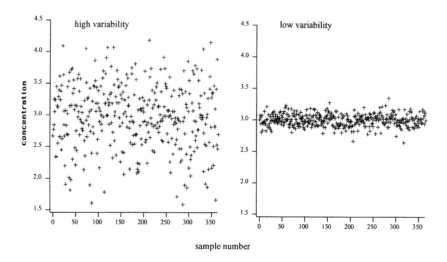

FIGURE 6-2 Effect of treated water variability on the concentration of a hypothetical contaminant.

ability situation would be undesirable, while the low-variability situation would be acceptable.

Multiple-barrier systems tend to produce less variability in the levels of contaminants that pass through the system. For instance, while single- and dual-process systems may pose similar mean levels of risk for contaminant breakthrough, dual-process systems greatly reduce the chance of high levels of contaminant breakthrough. This reduced variability makes multiple-barrier systems preferable to single-barrier systems. Realistic plant designs therefore contain many barriers, with the performance characteristics of various barriers carefully considered so that they complement each other.

Conceptually, the reliability of the overall treatment train can be approximated using event chain analysis, which has been formulated as a systematic technique for some other types of engineered systems (Kumamoto and Henley, 1996). The application of event chain analysis to a portion of a hypothetical reclamation facility is indicated in Figure 6-3, following Keeney et al. (1978).

In most cases, barriers neither perform perfectly nor fail completely, but rather have performance distributions over a broad continuum. A more precise method of reliability assessment than event chain analysis would need to use statistical methods to reflect the performance distributions of each barrier. Such statistical analysis, known as consequence-frequency assessment, has been previously used in microbial risk assess-

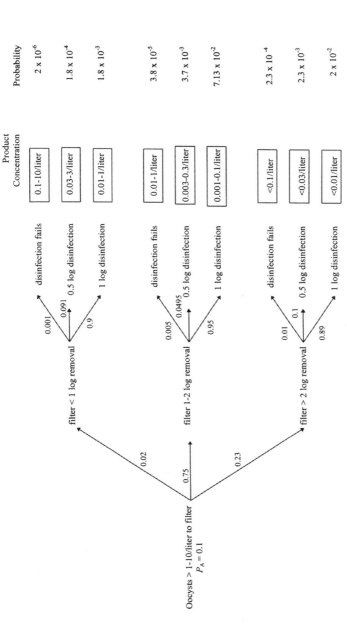

FIGURE 6-3 Schematic event tree analysis, as applied to a hypothetical water reclamation facility. Sets of arrows branching from a common point indicate alternative occurrences. Numbers on the arrows indicate occurrence probabilities. At the extreme right (in boxes) are the final occurrence magnitudes—here given as the microorganism density in the final product water (after filtration and disinfection) and probabilities of occurrence. The initiating event (with a probability of 0.1) is the occurrence of a microbial concentration of 1 to 10 *Cryptosporidium* oocysts per liter in a sand filter influent.

ment for evaluating the probability of contaminant breakthrough for individual elements along a process connecting a source of microorganisms to a potential receptor (Gerba and Haas, 1988; Haas et al., 1993; Regli et al., 1991; Rose et al., 1991). Appendix A provides a detailed example of the use of this method to assess the reliability of an advanced wastewater treatment facility.

PERFORMANCE EVALUATION, MONITORING, AND RESPONSE

One of the most important uses of reliability analysis is to determine critical processes that must be kept under tight control to limit the probability of high levels of contaminant exposure. Operational variables that provide early warning of failures in water treatment processes should be identified and incorporated into an ongoing monitoring-and-control strategy. These variables, which will be termed "sentinel parameters," should ideally be readily measurable on a rapid (even instantaneous) basis and should correlate well with high contaminant breakthrough. They need not be of particular health or environmental interest in and of themselves.

Sentinel Parameters

Appropriate sentinel parameters will depend upon the contaminant of interest and the particular process. The following are potential sentinel parameters for microbial breakthrough for some unit processes:

- For lime treatment and clarification, pH and turbidity can serve as sentinel parameters.
- For filtration, elevated particle counts or turbidity can signal problems.
- For reverse osmosis, sentinel parameters could include conductivity, chloride levels, total organic carbon, or transmembrane pressure.
- For ozonation, the dissolved ozone residual or UV_{254} could serve as sentinel parameters.

The use of sentinel parameters, combined with a sufficiently large environmental buffer or alternatives for diversion, might permit rapid operational adjustments to prevent microbial or other contaminant exposure of the population. The application of sentinel parameter monitoring to water reclamation facilities is analogous to the application of HACCP (hazard analysis and critical control points) methods to the food processing industry (Havelaar, 1994; Jay, 1992; Notermans et al., 1994).

The concept of sentinel parameters should be separated from the measurement of particular constituents that have direct health or regulatory interest. Sentinel parameters should be capable of being rapidly measured with respect to the rate at which underlying process fluctuations are anticipated and should be associated directly with potential contaminant breakthrough. Parameters measured for compliance purposes may require more time-consuming sample collection, preparation, and analysis procedures. The term "sentinel" is also used in contrast to "surrogate." Surrogate parameters are those whose measurement is intended to serve as a particular flag for the potential presence of a contaminant of health concern. Sentinel parameters are those designed to serve as a flag for the potential deterioration in process performance. While it is possible that some sentinel parameters may be surrogate parameters, and vice versa, this is not necessary.

Monitoring and Response

To ensure safety, every treatment facility should have in place rigorous monitoring and control systems to detect and correct lapses in performance.

Monitoring

The role of monitoring in a water reclamation plant, as in any water treatment plant, is to verify that the routine operational characteristics are in fact achieving the intended objectives. Typical routine monitoring parameters might include total organic carbon (TOC), nitrogen, phosphorus, coliforms, phage, *Clostridia*, and plate counts (see Chapters 2 and 3). Less routine parameters used for monitoring might include viruses, protozoa, pesticides, total organic halides, and heavy metals. Special studies might also use toxicological responses involving appropriate assays (see Chapter 5). Ideally, the entire battery of monitoring should be able to confirm that the routine operation of the facility and its design are in control.

Advances in analytical capability should make new monitoring tools and analytes available. However, as in any other environmental application, the development of more sensitive or selective methods does not necessarily mean that they should be employed on a routine basis. Instead, such developments in methodology should be used to verify that plant design and operation continue to protect public health. If necessary, special (nonroutine) studies might be necessary to develop new sentinel parameters for incorporation into the plant operational strategy.

Strategies for Monitoring and Control

A strategy for monitoring and controlling an indirect potable reuse system can include (1) continuous monitoring, (2) routine monitoring, and (3) ad hoc monitoring.

Continuous monitoring uses sentinel parameters that are directly relevant to the control of the process and for which reliable instruments for continuous monitoring are available. Monitoring must be at sampling locations that are relevant to process control. Continuous monitoring should always be carefully designed to allow sufficient time for blending and chemical reactions upstream of the point of sampling. When continuous monitoring is used directly in process control, the design should also consider lag times between the point where process changes are made and the point where samples are taken, as well as the response time for equipment conducting sample analysis and equipment involved in making process changes. Reliable instruments are essential.

Routine monitoring refers to regularly scheduled sampling and analysis that support both operation and regulatory compliance. For operational support, samples of sentinel parameters are generally obtained by grab samples and immediately analyzed on-site to help with hour-to-hour decision making. Samples taken for routine monitoring to support regulatory compliance are flow-proportioned composite samples taken on-site but analyzed by a certified drinking water laboratory. A routine monitoring program should comply with all regulatory monitoring requirements, in terms of both sampling frequency and analytes selected.

Ad hoc monitoring refers to special sampling and analysis designed to investigate process performance or to demonstrate the removal of special analytes that are not part of a permanent, routine monitoring program. Examples of ad hoc monitoring for process performance would be examining the profile of TOC removal throughout the process train or conducting special seeding challenges to demonstrate the removal of a particular contaminant in a particular unit process. Such ad hoc programs may use continuous, composite, or grab samples as appropriate.

The removal of special analytes, such as viruses, might be evaluated by a special ad hoc monitoring program during the first year of operation—an analysis that might be too expensive for long-term routine monitoring. An ad hoc monitoring program may include an extensive list of synthetic organic chemicals in order to determine those that should be included in future routine monitoring. Another example might be an ad hoc program to characterize the components of the organic carbon that still persist in the advanced wastewater treatment (AWT) product water.

Response

When a sentinel parameter monitoring program shows lapses in performance, appropriate actions should be taken to correct the situation. The appropriate action will depend upon the particular plant design. Ideally, the design should allow the operator to respond to a variety of circumstances. Such design flexibility might include extra pump capacity with reconfigurable piping and intermediate storage capacity, such as equalization and surge tanks.

Particular actions taken in response to process excursions might include the following:

- increasing the number of parallel process modules (filters, membranes, etc.) in service;
- removing and regenerating individual parallel process modules;
- adding supplemental chemicals or increasing chemical dosing;
- recirculating product water to reclamation plant headworks; or
- diverting product flow to an outfall, or to a nonpotable use, rather than into the potable supply stream.

Diversions should be implementable from all points within the reclamation facility, as shown in Figure 6-4 in the conceptual design for the San Diego Reclamation Facility (Bernados, 1996). In addition, process excursions might be used to trigger public health responses such as additional population surveillance, notification of sensitive subpopulations, or provision of emergency alternative supplies. The nature of any public health response will depend on the duration and magnitude of process excursions and the delay time provided by the environmental buffer and by finished water storage.

The existence of finished water storage (including perhaps storage of injected product water into an aquifer prior to use) either within or following a reclamation plant provides increased flexibility for operations. Such storage is beneficial in at least two respects. First, the additional holding time further removes contaminants by processes such as natural decay and biological and chemical reaction. Second, the storage may act to buffer fluctuations in quality so that short-lived transient spikes in undesirable contaminants are smoothed out prior to being sent for consumption. Such finished water storage may help to mitigate the urgency of process excursions. Conversely, the absence of such storage makes it more vital to continuously monitor processes and attend to deviant conditions. There is thus a trade-off between the stringency of reliability that may be needed in treatment and the size of the finished water storage area. Site-specific work will be necessary to quantify these trade-offs.

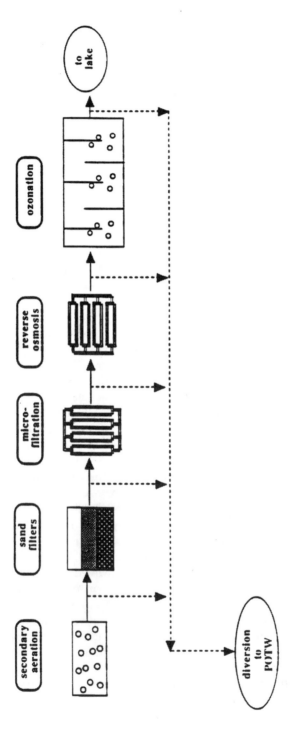

FIGURE 6-4 Diagram of conceptual design of San Diego Reclamation Plant (after Bernados, 1996). Dashed lines indicate potential emergency diversion points. NOTE: POTW = publicly owned treatment works. SOURCE: Reprinted from AWWA Water Reuse Conference Proceedings (San Diego, California, February 25-28, 1996), by permission, © 1996, American Water Works Association.

REDUCING RISK FROM UNIDENTIFIED TRACE ORGANIC CONTAMINANTS

The organic matter in the water produced by an aerobic biological wastewater treatment plant consists of essentially two components: (1) synthetic organic chemicals (SOCs) of anthropogenic origin, as defined by the Environmental Protection Agency (EPA) Office of Drinking Water, and (2) natural organic matter (NOM), which is mostly an ill-defined set of compounds generated through microbial metabolism (see discussion in Chapter 2). Reclaimed water generally holds higher concentrations of both types of components than conventional drinking water supplies do. However, in both reclaimed water and conventional drinking water supplies, the SOCs may not be distinguishable from the NOM. And either water may hold significant quantities of chemicals having potent toxicological properties (e.g., endocrine disrupters, pharmaceutical agents or metabolites, or hormones).

The state of our knowledge concerning SOCs has improved since the 1982 National Research Council report *Quality Criteria for Water Reuse* (NRC, 1982). Extensive lists of organic compounds have been generated based on known ground water contaminants, the priority and toxic pollutant lists of the Clean Water Act, and the Drinking Water Priority List (53 Federal Register 1892, January 1988, updated 56 Federal Register 1470, January 1991).

Using these lists as a guide, analytical methods have been developed to detect most of the organic compounds of concern in industrial discharges, in raw sewage and treated sewage, and in drinking water supplies. The discharge of priority pollutants to the sewer and to the environment has been regulated and enforced by monitoring requirements. As a result of these efforts, more is known about the anthropogenic chemicals in sewage today, and discharge of these compounds to the nation's sewers is more tightly controlled than was the case in 1982. While concern about chemical risk from reclaimed water remains, the potential for risk management is greater today than in 1982.

The largely uncharacterized organic matter in natural water, in conventionally treated wastewater, and in reclaimed water consists of large organic macromolecules usually identifiable in only the broadest way (e.g., by functional groups, molecular weight, aromacity, or acid/base solubility). As explained in Chapter 2, we have a poor understanding of the differences, if any, between NOM generated in biological sewage treatment processes and NOM generated in a conventional watershed. A credible argument can be made that the same biochemical processes produce the organic matter in both circumstances, and their chemistry is likely to be much the same, with only subtle differences in molecular weight and functional groups.

As in conventional drinking water supplies, the toxicity of uncharacterized organic materials in reclaimed water may be of less concern than the toxicity of by-products produced when these materials are transformed by disinfection processes. Some experimental work with concentrates of organic chemicals from reclaimed water has shown no toxic effects on animals (for example, the Tampa and Denver studies described in Chapter 5). Nevertheless, as a society, we have less experience with exposure to wastewater-derived organic matter than we do with the NOM in conventional water supplies. Therefore, the concentration of wastewater-derived organic matter should be minimized as part of any prudent potable reuse plan.

To address SOCs, the EPA should develop a priority list of contaminants of public health significance that are known or anticipated to occur in sewage. This list should then be used by utilities planning potable reuse as they manage their industrial pretreatment programs so that the introduction of these compounds into the sewer is regulated and monitored. Finally, potable reuse operations should include a program to monitor for these chemicals in the AWT effluent. Chemicals that occasionally occur at measurable concentration should be monitored at greater frequency than those are not typically detected.

Every community considering potable reuse should carefully review its industrial pretreatment program to ensure that this program serves sufficiently as the principal barrier to SOCs. The industrial pretreatment program must accomplish three essential goals: (1) identify the potentially toxic compounds used or produced by industry, (2) establish regulations to prevent their discharge to the sewer, and (3) establish monitoring to ensure that discharge does not occur and to detect the presence of these compounds in the wastewater should controls fail. This same monitoring program can serve as a basis for designing the chemical monitoring system for the reclaimed water.

To address health risks associated with the contribution of NOM to the potential for disinfection by-product formation, the TOC of the reclaimed water must be reduced to the lowest feasible level, as recommended in Chapter 2. The additional barriers in indirect potable reuse, such as dilution, soil-aquifer treatment, and long retention times in surface reservoirs and/or ground water aquifers, will contribute to the overall reduction of organic carbon of wastewater origin to the water supply. As TOC of wastewater origin diminishes, so do the health concerns associated with it. In principle there comes a point where these concerns are less important than other concerns already being addressed by the current drinking water regulations and their associated monitoring requirements. When the wastewater-derived TOC is below this level, no special toxicological monitoring should be required. When the wastewater-derived TOC is above this level, public safety should be protected with continuous toxicological monitoring using *in vivo* systems (see Chapter 5).

Although establishing such a TOC level appears to be a legitimate risk management strategy, there is no scientific basis for determining what that level should be. The committee believes this judgment should be made by local regulators, integrating all the information they have available to them concerning a specific project.

PUBLIC HEALTH SURVEILLANCE

Public health surveillance programs are essential components of a community-wide strategy to provide early warning of possible health problems. Adequate disease surveillance requires continuing scrutiny of all aspects of occurrence and spread of a disease that are pertinent to effective control.

The most comprehensive and internationally accepted definition of public health surveillance is that found in the American Public Health Association report entitled *Control of Communicable Diseases Manual* (Benenson, 1995). That report calls for the systematic collection and evaluation of

1. morbidity and mortality reports;
2. special reports of field investigations of epidemics and individual cases;
3. isolation and identification of infectious agents by laboratories;
4. data concerning the availability, use, and untoward effects of vaccines and toxins, immune globulins, insecticides, and other substances used in disease control;
5. information regarding immunity levels in segments of the population; and
6. other relevant epidemiologic data.

A report summarizing the above data should be prepared and distributed to all of those involved in public health protection. This procedure applies to all jurisdictional levels of public health protection, from local to international (Benenson, 1995).

This definition of surveillance of disease is distinct from health surveillance of specific persons, which is a form of public health quarantine. Any surveillance programs should be tailored to the needs of its community, and not all of the elements of the definition above necessarily apply to populations exposed to potable water supplies augmented with re-claimed water. Figure 6-5 diagrams relationships between the types of public health surveillance (hazard surveillance, exposure surveillance, and outcome surveillance) and the corresponding process by which an environmental agent produces an adverse effect (Thacker et al., 1996).

Surveillance is distinct from epidemiological studies in that surveillance is an ongoing public health program analogous to continuous monitoring. Epidemiology, in contrast, is an investigative activity on the patterns of disease occurrence in particular human populations (or study groups) to identify the factors that influence these patterns of disease. The investigative activity may be descriptive, analytical, or experimental in form. Findings identified in surveillance programs may generate hypotheses that could be tested by epidemiological studies.

The ability of a surveillance system to provide early warning of a possible health problem in the community depends on its sensitivity or threshold of detection. The sensitivity varies depending on the type of surveillance (passive vs. active) and the resources available to maintain the system. In most waterborne diseases, there is normally a low level of sporadic cases in the population, known as "endemic" illness. As the number of exposed individuals increases, the number of recognized cases will rise. This may occur rapidly for an infectious disease with a short incubation period or slowly for a chronic disease with a long or variable latency period. An increase in the illness rate will be recognized by the surveillance system as an outbreak when the disease rate exceeds the threshold of detection. The outbreak event should trigger an epidemiologic investigation to determine (1) whether the reported increases in disease are real or an artifact of the reporting or detection methods; (2) whether the cases are related, as by geographic proximity or common exposure; (3) whether the diseases may be related to any changes in measured water quality parameters; and (4) how the outbreak can be controlled and further disease prevented.

The Uses of Surveillance

Public health surveillance can be put to many uses, including the following (Teutsch and Churchill, 1994):

- developing quantitative estimates of the magnitude of a health problem;
- portraying the natural history of disease;
- detecting epidemics;
- documenting the distribution and spread of a health event;
- facilitating epidemiologic and laboratory research;

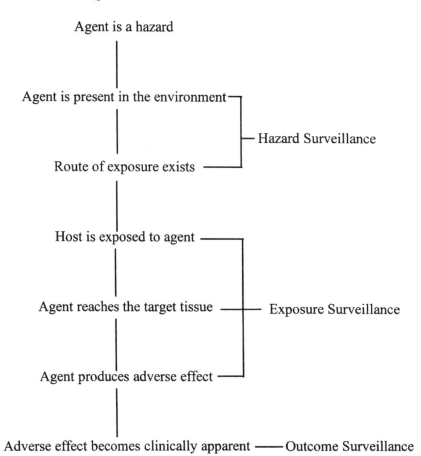

FIGURE 6-5 The process by which an environmental agent produces an adverse effect and the corresponding types of public health surveillance necessary to monitor that effect. SOURCE: Reprinted, with permission, from Thacker et al., 1996. © 1996 by American Journal of Public Health.

- testing hypotheses;
- evaluating control and prevention measures;
- monitoring changes in infectious agents;
- detecting changes in health practice; and
- planning.

Surveillance may be used to identify and track waterborne health hazards even when the water reclamation facility is operating in accordance with applicable regulations and within generally accepted standards. In a report by Tilden et al. (1991), for example, dry fermented salami was shown to serve as a vehicle of transmission for O157:H7 strains of *Escherichia coli*, even when all the food production methods used complied with existing regulations and recommended good manufacturing practices. In this case, Tilden et al. noted that "surveillance . . . provides the ultimate feedback on the efficacy of the standard industry safety plans."

Public Health Surveillance and Water Reclamation

Recent developments in the application of public surveillance to drinking water uses in general are also relevant to reclaimed water projects. The 1993 outbreak of cryptosporidiosis affecting more than 400,000 persons in Milwaukee stimulated a great deal of interest in the role of public health surveillance in the early detection of harmful agents in municipal water supplies.

The Centers for Disease Control and Prevention (CDC) held a workshop in September 1994 to assess the public health threat associated with waterborne cryptosporidiosis (CDC, 1994). One of the workshop's major objectives was "to identify surveillance systems . . . for assessing the public health importance of low levels of *Cryptosporidium* oocysts or elevated turbidity in public drinking water." The major observations and recommendation published in the report of that workshop are as follows:

• Local public health officials should consider developing one or more surveillance systems to establish baseline data on the occurrence of cryptosporidiosis among residents of their community and to identify potential sources of infection.
• No single surveillance strategy is appropriate for all locations; therefore, communities should select a method that meets local needs and is most compatible with existing disease surveillance systems.
• Cryptosporidiosis should be reportable to the CDC.
• Sales of antidiarrheal medications should be monitored. The development of an information exchange system between local pharmacists and state or local public health officials is a cost-effective and timely way to detect increases in diarrheal illness in some counties.
• Logs maintained by health maintenance organizations and hospitals should be monitored for diarrheal illness.
• Information entered promptly into a computerized database can effectively monitor both complaints of diarrhea and severity of gastrointestinal disease in a community.
• Incidence of diarrhea in nursing homes should be monitored. Diarrheal illness rates in residents of nursing homes that use municipal

drinking water can be compared with illness rates in residents of other nursing homes in the same community that use a different water source (e.g., private well water).

• Laboratories should routinely test for *Cryptosporidium* in stool specimens. Most laboratories presently do not look for *Cryptosporidium* in specimens submitted for routine parasitologic examination.

• Tap water should be monitored for *Cryptosporidium*. This information will allow researchers to see how often an increase in diarrheal disease or *Cryptosporidium* diagnosis occurs during the first week to two weeks after oocysts are found in drinking water.

• Local, state, and national public health agencies should cooperatively initiate and develop surveillance systems to assess the public health significance of low levels of *Cryptosporidium* oocysts in public drinking water.

A similar investigation was undertaken by the New York Department of Environmental Protection (DEP). DEP commissioned an advisory panel on waterborne disease assessment to determine whether New York City's active disease surveillance program (1) was adequate to detect a waterborne disease outbreak; (2) could provide sufficient information to assess the endemic rates of cryptosporidiosis and giardiasis; and (3) could provide sufficient information to assess endemic waterborne disease risks. The panel was also asked to assess whether an epidemiologic study or studies could be designed to address the question of increased risk of enteric illness or infection associated with the consumption of tap water.

The approach of the New York advisory panel on waterborne disease assessment was similar to that of the CDC workshop, and its conclusions are likewise relevant to the issue of augmenting potable water supplies with reclaimed water. The panel made five major recommendations:

1. Designate a waterborne disease coordinator. The most important aspect of a successful waterborne disease surveillance program is the designation of an individual who is specifically responsible for developing, supervising, and coordinating all aspects of the program.

2. Initiate rapid reporting and analysis of disease surveillance data. Early detection of a waterborne outbreak requires that disease surveillance information be quickly transmitted and analyzed.

3. Initiate special waterborne disease surveillance studies. Studies should include (1) surveillance of diarrheal illness and selected infectious diseases in populations using managed care programs (e.g., health maintenance organizations), in selected emergency treatment facilities, and in

sentinel populations residing in nursing and/or retirement homes; and (2) monitoring of the sales or use of medications for diarrheal illness.

4. Improve reporting of cryptosporidiosis. The reporting of cryptosporidiosis can be improved by educating physicians and health care providers about the disease and encouraging laboratories to examine stool specimens for *Cryptosporidium*.

5. Institute an annual evaluation of the waterborne disease surveillance program to make sure it is effective for the detection and early recognition of waterborne outbreaks or emerging pathogens.

The designation of a waterborne disease coordinator is not the norm, as was documented by Frost et al. (1995) in their report of a survey of state and territorial programs. Frost et al. found that in 49 states and 3 territories surveyed, fewer than half of the programs had designated a waterborne disease coordinator. Of these coordinators, only 24 percent had training or work experience in water treatment and only 28 percent met regularly with drinking water regulatory staff. Only 46 percent of states and territories surveyed indicated that an individual from the drinking water regulatory staff had been designated to assist in investigation of outbreaks, and almost a third of those designated individuals could not be named by the epidemiology program staff. No state used computerized illness data from health maintenance organizations or the Indian Health Service as a waterborne disease surveillance tool. With fewer than half of the state/territorial epidemiology programs having a designated coordinator for waterborne disease outbreaks and with few of these coordinators having training or experience in drinking water treatment or maintaining contact with the state water treatment specialists, the timeliness of initiation and the effectiveness of outbreak investigations may be compromised.

Strengths and Limitations of Surveillance Systems

The basic structures for public health surveillance are already in place in most communities. Mandatory disease reporting systems have been established by statute or regulations in all state and many local jurisdictions, and all states require that physicians report cases of specified diseases to the appropriate state or local health department. The diseases that are reportable, however, vary from state to state. Reporting agencies such as physicians and laboratories are repeatedly reminded of their obligations in this regard (Rutherford, 1992), and lists of diseases to be reported are frequently updated. In California, for example, *Escherichia coli* O157:H7 and any "waterborne disease" were added to the California

Code of Regulations and were required to be reported by telephone. The list already contained cryptosporidiosis and giardiasis as well as the more classical salmonellosis and shigellosis (Ross, 1995).

However, as stated by Teutsch and Churchill (1994), "Under-reporting is a consistent and well-characterized problem of notifiable-disease reporting systems. In the United States, estimates of completeness of reporting range from 6 percent to 90 percent for many of the common notifiable diseases." Many factors contribute to this lack of complete reporting, but they are generally well known and can be addressed by positive action if the will to do so exists. As Teutsch and Churchill (1994) note,

> Some approaches that appear to be successful include (a) providing physicians with feedback on the health department's disposition of individual cases; (b) matching laboratory reports with physicians' reports, and for those cases reported only by laboratories, notifying physicians that a specific case should have been reported to the health department; and (c) conducting in-person site visits to review reporting procedures.

The ultimate purpose of surveillance systems is to prevent or control the occurrence of adverse health events associated with the ingestion of drinking water sources augmented with reclaimed water. Therefore, any such surveillance system must be jointly planned and operated not only by those who collect the data, but by those who would use these data in conjunction with other monitoring processes to ensure the quality of the water delivered to the consumer. As a minimum, this would include the health, water, and wastewater departments. Essential to the effective functioning of such a system is the identification of key individuals in each agency who would plan, coordinate, rehearse, and communicate frequently. This might be tied in with the community's general emergency response plan. It would also be appropriate to include and advise interested consumer groups of the surveillance plan and its purpose.

The strength of the surveillance system is directly proportional to the degree to which this coordinated joint system is effective, and it will be limited to the degree that coordination is lacking.

OPERATOR TRAINING AND CERTIFICATION

Proper operation of an advanced water treatment plant intended to improve potability of wastewater requires special training. Neither the conventional wastewater treatment operator nor the conventional water treatment operator gets sufficient training or experience in physicochemical treatment or in public health microbiology to serve the needs that must be met at an AWT plant.

Such training should include at a minimum the principles of operation of processes for coagulation with ferric chloride or lime (especially high-lime treatment), granular media filtration, membrane filtration, reverse osmosis, air stripping, ion exchange, advanced oxidation, and high-level disinfection with free chlorine, ozone, UV light, and other means.

Training should include courses on the various microbial pathogens and indicators. Operators should be familiar with many of the most important microbial organisms, the diseases they cause, the symptoms of those diseases, the likely density of the organisms in wastewater, and the relative effectiveness of the various treatment processes in removing each one. Operators should also be generally familiar with the procedures for isolating these organisms from drinking water as well as some of strengths and weaknesses of the various analytic techniques.

CONCLUSIONS AND RECOMMENDATIONS

The safe, reliable operation of a potable reuse water system depends both on well-designed treatment trains that provide redundant safety measures, or "multiple barriers," and on monitoring efforts designed to detect variations in system operation as well as any signs of contaminant breakthrough in the system. Such duplicative barriers and monitoring efforts are essential to reducing, detecting, and mitigating any weaknesses or lapses in the system's safety performance.

To provide these margins of safety, the committee recommends the following:

• **Potable water reuse systems should employ independent multiple barriers to contaminants, and each barrier should be examined separately for its efficacy for removal of each contaminant.** Further, the cumulative capability of all barriers to accomplish removal should be evaluated, and this evaluation should consider the levels of the contaminant in the source water, the expected health effect associated with the contaminant, the goals that have been set for the potable supply, and any additional factors of safety.

• **The multiple barriers for microbiological contaminants should be more robust than those for many other forms of contamination, due to the acute danger such contaminants pose at high doses even for short time periods.** Where reclaimed water is used to augment natural supplies or where source water cannot be protected from upstream discharges of water with impaired quality, the importance of the barriers at the water treatment plant or wastewater reclamation plant is correspondingly increased.

• **Because the performance of wastewater treatment processes may vary considerably from time to time, such systems should employ** quantitative reliability assessments to gauge the probability of contaminant breakthrough among individual unit processes. Such a quantitative approach can be combined with dose-response assessments to better ascertain the likelihood of a risk of infection or illness of a given magnitude. Sentinel parameters, which are readily measurable on a rapid (even instantaneous) basis and which correlate well with high contaminant breakthrough, should be used for monitoring critical processes that must be kept under tight control.

• **Utilities using surface waters or aquifers as environmental buffers should take care to prevent "short-circuiting," by which influent treated wastewater either fails to mix with the ambient water fully or moves through the system to the drinking water intake faster than expected.** In addition, the buffer's expected retention time should be long enough to give the buffer time to provide additional contaminant removal. California has proposed regulations requiring retention times of 6 to 12 months before treated wastewater can be reused as a potable water source; these storage times seem appropriate for enabling the development and execution of alternative actions.

• **Risk management strategies should be used to reduce the risk from the wide variety of synthetic organic chemicals that may be present in municipal wastewater and consequently in reclaimed water.** Utilities involved with planning for potable reuse should implement a stringent industrial pretreatment and pollutant source control program that not only includes existing priority pollutants, but considers additional contaminants of public health significance that are known or anticipated to occur in sewage. Guidelines for developing lists of wastewater-derived SOCs should be prepared by the Environmental Protection Agency and modified for local use. Finally, potable reuse operations should include a program to monitor for these chemicals in the AWT effluent, tracking those that occasionally occur at measurable concentration with greater frequency than those that do not.

• **Potable reuse operations should have alternative means for disposing of the reclaimed water in the event that it does not meet required standards.** Such alternative disposal routes protect the environmental buffer from contamination.

• **Every water agency using reclaimed waters as drinking water should implement well-coordinated public health surveillance systems to document and possibly provide early warning of any adverse health events associated with exposure to reclaimed water.** Such surveillance data should be interpreted with care, because there are many sources of exposure for all diseases categorized as "waterborne." (That is, suggested elevated levels of disease may be due to other transmission routes and not related to drinking water.) There is little scientific documentation of the degree to which most specific diseases are due to exposure to water. It is important to have good communication and cooperation between health authorities and water utilities to quickly recognize and act upon any health problems that may be related to water supply. Any such surveillance system must be jointly planned and operated by the health, water, and wastewater departments and should take advantage of recommendations made in conjunction with recent public health workshops on the subject sponsored by the Centers for Disease Control and Prevention. Essential to the effective functioning of such a system is the identification of key individuals in each agency who would coordinate planning and rehearse emergency procedures. Further, appropriate interested consumer groups should be involved and informed as to the public health surveillance plan and its purpose.

• **Operators of water reclamation facilities should receive adequate training that should include the principles of operation of advanced treatment processes, a knowledge of pathogenic organisms, and the relative effectiveness of the various treatment processes in reducing pathogen concentrations.** Operators of such facilities need training beyond that typically provided to operators of conventional water and wastewater treatment systems.

REFERENCES

American Water Works Association. Organisms in Water Committee. 1987. Committee report: microbiological considerations for drinking water regulation revisions. Journal of the American Water Works Association 79(5): 81-88.

Benenson, A. S., ed. 1995. Control of Communicable Diseases Manual, 16th ed. Washington, D.C.: American Public Health Association.

Bernados, B. 1996. The Importance of Reliability in Potable Reuse. AWWA Water Reuse Conference. San Diego, Calif., February 25-28.

Centers for Disease Control and Prevention (CDC). 1995. Morbidity and Mortality Weekly Report 44 (No. RR-6:1-19). Atlanta, Ga.: CDC.

Ellis, R. H. 1995. Seattle Water Department: Cedar River Surface Water Treatment Rule Compliance Project. Seattle, Wash.: Seattle Water Department.

Englehardt, J. D. 1995. Predicting incident size from limited information. Journal of Environmental Engineering 121(6): 455-464.

Fischer, H., J. List, R. Koh, H. Imberger, and N. Brooks. 1979. Mixing in Inland and Coastal Water. New York: Academic Press.

Flow Science for Montgomery Watson and City of San Diego. 1995. Vicente Reclamation Project: Results of Tracer Studies. Pasadena, Calif.: Flow Science, Inc.

Frost, F. J., R. L. Calderon, and G. F. Craun. 1995. Waterborne disease surveillance: Findings of survey of state and territorial epidemiology programs. Environmental Health Dec.

Gerba, C. P., and C. N. Haas. 1988. Assessment of risks associated with enteric viruses in contaminated drinking water. ASTM Special Technical Publication 976: 489-494.

Haas, C. N. 1997. Importance of distributional form in characterizing inputs to Monte Carlo risk assessment. Risk Analysis 17(1):107-113.

Haas, C. N., J. B. Rose, C. P. Gerba, and S. Regli. 1993. Risk assessment of virus in drinking water. Risk Analysis 13(5): 545-552.

Havelaar, A. 1994. Application of HACCP to drinking water supply. Food Control 5(3):145-152.

Herwaldt, B. L., et al. 1991. Waterborne-disease outbreaks, 1989-1990. Morbidity and Mortality Weekly Report 40(SS-3):1-21.

James M. Montgomery Consulting Engineers Inc. 1985. Water Treatment Principles and Design. New York: John Wiley.

Jay, J. M. 1992. Microbiological food safety. Critical Reviews in Food Science and Nutrition 31(3):177-190.

Keeney, R., et al. 1978. Assessing the risk of an LNG terminal. Technology Review 81(1): 64-78.

Kumamoto, H., and E. Henley. 1996. Probabilistic Risk Assessment and Management for Engineers and Scientists. New York: Institute of Electrical and Electronics Engineers, Inc.

Lippy, E., and S. Waltrip. 1984. Waterborne disease outbreaks—1946-1980: a thirty-five-year perspective. Journal of the American Water Works Association 76(2):60-67.

National Research Council (NRC). 1982. Quality Criteria for Water Reuse. Washington, D.C.: National Academy Press.

National Research Council (NRC). 1994. Ground Water Recharge Using Waters of Impaired Quality. Washington, D.C.: National Academy Press.

New York City's Advisory Panel on Waterborne Diseases Assessment. 1994. Report of New York City's Panel on Waterborne Disease Assessment. The New York City Department of Environmental Protection.

Notermans, S., et al. 1994. The HACCP concept: specification of criteria using quantitative risk assessment. Food Microbiology 11(5):397-408.

Raphael, J. M. 1962. Prediction of temperature in rivers and reservoirs. J. Power Div. Am. Soc. Civ. Engr. 99:475-1496.

Regli, S., et al. 1991. Modeling risk for pathogens in drinking water. Journal of the American Water Works Association 83(11):76-84.

Rose, J. B., C. N. Haas, and S. Regli. 1991. Risk assessment and the control of waterborne giardiasis. American Journal of Public Health 81:709-713.

Ross, R. K. 1995. Additional diseases to be reported. Physician's Bulletin. June (No. 398). County of San Diego.

Rutherford, G. W. 1992. Why-when-how to report communicable diseases. Medical Board of California Action Report 45(May):24.

Smith, R. L. 1994. Use of Monte Carlo simulation for human exposure assessment at a Superfund site. Risk Analysis 14(4):433-439.

Sontheimer, H. 1980. Experience with riverbank filtration along the Rhine River. Journal of the American Water Works Association 72(7):386.

Teutsh, S. M., and R. E. Churchill. 1994. Principles of Public Health Surveillance. New York: Oxford University Press.

Thacker, S. B., D. F. Stroup, R. G. Parrish, and H. A. Anderson. 1996. Surveillance in environmental public health: issues, systems, and sources. Am. J. Public Health 1996: 633-638.

Tilden, J., Jr., W. Young, A. M. McNamara, C. Custer, B. Boesel, M. A. Lambert-Fair, J. Majkowski, D. Vugia, S. B. Werner, J. Hollingsworth, and J. G. Morris, Jr. 1991. A new route of transmission for *Escherichia coli*: infection from dry fermented salami. Am J. Public Health 86:1076-1077.

Velz, C. 1970. Applied Stream Sanitation. New York: John Wiley.

A

Use of Consequence-Frequency Assessment to Evaluate Performance of an Advanced Wastewater Treatment Facility

The consequence-frequency assessment methodology (see Chapter 6) can be illustrated by example. For the purpose of this discussion, only advanced wastewater treatment facilities subsequent to secondary wastewater treatment are considered. For acute contaminants (such as microorganisms), the influent stream to such a facility may be highly variable. For example, at Water Factory 21, the range in influent virus levels was more than two orders of magnitude, and the distribution could be roughly described by a log normal distribution (see Figure A-1).

A given facility may have a sequence of processes individually capable of changing the microbial distribution. Figure A-2 depicts such a treatment in general terms. The concentrations of pollutants (e.g., viruses) at each stage of treatment are given as C_0, C_1, C_2, and so on. The connection between individual concentrations is given by a conditional distribution. For example, $F_2(C_2 \mid C_1)$ is the probability density of virus following filtration, given the virus concentration entering the filter. In other words, if the filter influent contains a virus concentration C_1 then the integral

$$\int_{C_2^a}^{C_2^b} F_2(C_2 \mid C_1)dC_2$$

gives the probability that the filter effluent will contain a virus concentration between C_2^a and C_2^b. Where the processes behave in a linear (first-order) fashion, this distribution may be written simply in terms of the

241

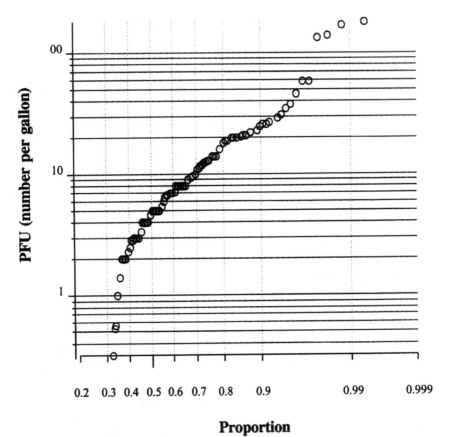

Proportion

FIGURE A-1 Influent (secondary effluent) virus concentrations on BGM cells at Water Factory 21. The x-axis is a normal probability scale. NOTE: BGM = Buffalo Green Monkey; PFU = plaque-forming units. SOURCE: Reprinted, with permission, from James Montgomery Consulting Engineers, 1979. ©1979 by James Montgomery Consulting Engineers.

ratio of the effluent to influent concentration; however the conditional framework (as shown in Figure A-2) provides a more general approach

Formally the probability distribution of the product concentrations may be evaluated as a multiple integral, which can (for the indicated process train) be written as (Stuart and Ord, 1987)

$$f_5(C_5) = \iiint \iint f_0(C_0) F_1 F_2 F_3 F_4 \, dC_0 \, dC_1 \, dC_2 \, dC_3 \, dC_4 \qquad (1)$$

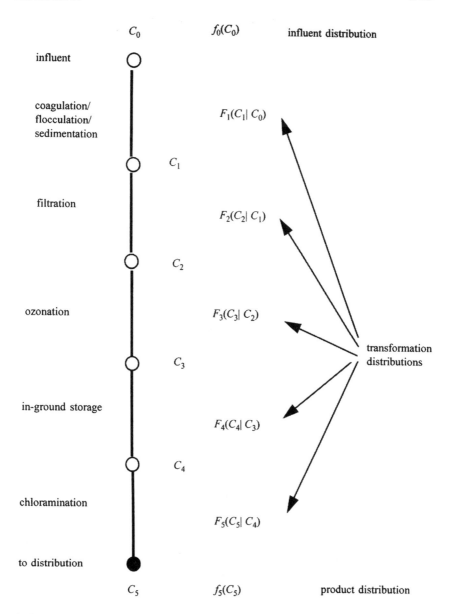

FIGURE A-2 Conceptual diagram of process train and reliability distributions.

For compactness, the arguments for the conditional distributions are omitted in this integral. Analytical evaluation of this integral may not be possible in many (perhaps most) cases. Therefore the problem must be approached numerically, and there are at least two numerical approaches to the problem. In one approach, the deterministic function connecting C_0 and C_5 is written:

$$C_5 = C_0 \left(\frac{C_1}{C_0} \right) \left(\frac{C_2}{C_1} \right) \left(\frac{C_3}{C_2} \right) \left(\frac{C_4}{C_3} \right) \left(\frac{C_5}{C_4} \right) \tag{2}$$

This can also be written as

$$\ln(C_5) = \ln(C_0) + \sum_{i=1}^{5} \ln\left(\frac{C_i}{C_{i-1}} \right) \tag{3}$$

If the moments (mean, variance, etc.) of the influent distribution (C_0) and the process removal (C_i/C_{i-1}) distributions, along with any correlations, are known, then the moments of the quantity $\ln(C_5)$ can be computed using propagation methods (Ku, 1966). However, this method is algebraically complex, and unless further assumptions are made, it cannot directly produce confidence distributions.

The Monte Carlo method is an alternative that is widely employed in risk assessment (Burmaster and Anderson, 1994; Finkel, 1990; Haas et al., 1993; Smith, 1994). In this approach, each of the distributions is sampled repeatedly, and the final concentration is computed for each set of random samples. By repeated sampling, the distribution of final concentrations is obtained.

To employ either of these methods, information about the transformation probability distributions must be obtained or assumed. For example, in the Water Factory 21 project (James M. Montgomery Consulting Engineers, 1979), paired samples of influent and clarified lime-treated water were analyzed for viruses. The resulting data are presented in the two panels of Figure A-3. It would appear that the removal in this process can be characterized by a log normal distribution of a removal efficiency (effluent/influent); however, the low number of data points does not provide a powerful test of alternative models. (For example, is removal efficiency dependent in a systematic manner on the influent viral loading or on other process descriptors such as pH?)

Obtaining information on process removal distributions is a data-intensive task. The actual data in Figure A-3 span from 10 to 90 percent cumulative probability. The implementation of a full Monte Carlo analy-

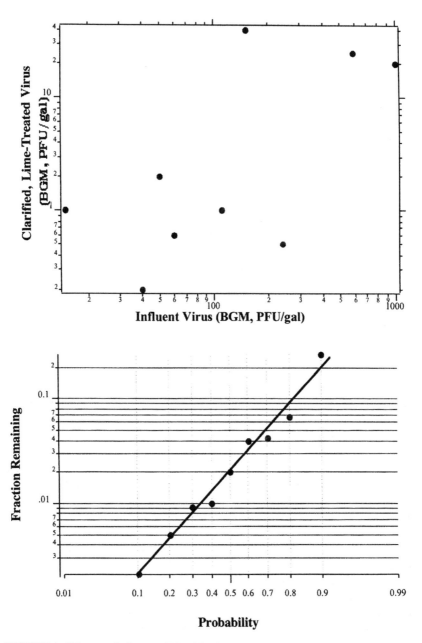

FIGURE A-3 Removal of virus (BGM) by lime treatment at Water Factory 21. The x-axis in the bottom panel is a normal probability scale. NOTE: BGM = Buffalo Green Monkey; PFU = plaque-forming units.

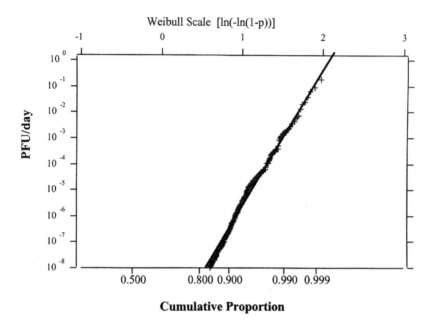

FIGURE A-4 Example of output of a Monte Carlo simulation. NOTE: PFU = plaque-forming units.

sis would require fitting this distribution to a particular form, which can then be used to extrapolate beyond the "tails" of the observed data. There are an infinite number of potential distributions. While the log normal has most frequently been used, the gamma and pareto may also have particular utility in the description of relatively rare failure distributions. The degree to which the choice of a particular distributional assumption influences the results of a Monte Carlo calculation has not been well studied, although preliminary results suggest that the choice of distributional form may not be overly crucial except in very highly skewed situations.

In principle, if the data on process performance distributions are available, then the probability distribution of a particular contaminant in the water to be ingested may be estimated. The final results may be presented as Figure A-4. Results from hypothetical data are plotted on a Weibull scale, which would result in a linear plot if the Weibull distribution provided a good fit to the data. From this figure one could estimate, for example, that there would be exposure to 10^{-3} plaque forming units per day more than 1 percent of the time. This information may be combined with a dose-response assessment to ascertain the likelihood of a risk of infection or illness of a given magnitude.

REFERENCES

Burmaster, D. E., and P. D. Anderson. 1994. Principles of good practice for the use of Monte Carlo techniques in human health and ecological risk assessment. Risk Analysis 14(4): 477-481.

Finkel, A. M. 1990. Confronting Uncertainty in Risk Management. Washington D.C.: Resources for the Future, Center for Risk Management.

Haas, C. N., J. B. Rose, C. P. Gerba, and S. Regli. 1993. Risk assessment of virus in drinking water. Risk Analysis 13(5):545-552.

James M. Montgomery Consulting Engineers. 1979. Water Factory 21 Virus Study, Orange County Water District. New York: John Wiley.

Ku, H. 1966. Notes on the use of propagation of error formulas. Journal of Research of the National Bureau of Standards—C: Engineering and Instrumentation 70C(4):263-273.

Smith, R. L. 1994. Use of Monte Carlo simulation for human exposure assessment at a Superfund site. Risk Analysis 14(4):433-439.

Stuart, A., and J. K. Ord. 1987. Kendall's Advanced Theory of Statistics. New York: Oxford University Press.

B

Biographies of Committee Members and Staff

COMMITTEE MEMBERS

JAMES CROOK, *Chair*, received his B.S. in civil engineering from the University of Massachusetts and his M.S. and Ph.D. in environmental engineering from the University of Cincinnati. He is director of water reuse for the firm Black & Veatch. He was previously with the California Department of Health Services, where he directed the department's water reclamation and reuse program. Dr. Crook has served on several water reuse advisory panels and has been an adviser to the National Sanitation Foundation, Pan American Health Organization, United Nations Development Programme, and U.S. Agency for International Development. He was the principal author of water reuse guidelines published by the U.S. Environmental Protection Agency and has assisted in the development of water reclamation and reuse criteria for several states.

RICHARD S. ENGELBRECHT, *Chair (through August 1996)*, received a B.A. from Indiana University and M.S. and Sc.D. degrees from the Massachusetts Institute of Technology. He was a professor of environmental engineering at the University of Illinois at Urbana-Champaign. Dr. Engelbrecht was one of the world's experts in the field of water pollution research and water quality control and was a member of the National Academy of Engineering. In 1986 he chaired the National Research Council's (NRC's) Committee on Recycling, Reuse, and Conservation in Water Management for Arid Areas, and in 1988-1990 he chaired the NRC's Committee to Review the U.S. Geological Survey's National Water Quality Assessment Pilot Program. He was also a member of the

248

NRC's Mexico City Water Supply (1995) and Quality Criteria for Water Reuse (1982) committees and was a founding member of the Water Science and Technology Board.

MARK M. BENJAMIN earned a B.S. in chemical engineering from Carnegie-Mellon University and an M.S. in chemical engineering and a Ph.D. in civil engineering from Stanford University. He is currently a professor in the Department of Civil Engineering at the University of Washington. His research interests include aquatic chemistry, the structure and reactivity of natural organic matter, treatment of toxic metals in industrial and municipal wastewater, and the chemistry of corrosion in water treatment systems.

RICHARD J. BULL received his B.S. in pharmacology from the University of Washington and his Ph.D. in pharmacology from the University of California, San Francisco. He spent 14 years in Cincinnati, where his last position was as director of the Toxicology and Microbiology Division of the Environmental Protection Agency's Health Effects Research Laboratory. He spent 10 years on the faculty of the College of Pharmacy at Washington State University (1984-1994). In 1994, he moved to Battelle's Pacific Northwest Division, where he works in the Pacific Northwest National Laboratory of the Department of Energy. His research has remained in the area of health hazards associated with drinking water from a variety of sources. His most recent research deals with mechanisms of adverse effects that are induced by the haloacetate group of disinfectant by-products. Dr. Bull is an active member of several societies and has served on several NRC committees, including the Subcommittee on Disinfectants and the Committee on Recycling, Reuse, and Conservation in Water Management for Arid Areas.

BRUCE A. FOWLER earned a B.S. in fisheries/marine biology from the University of Washington and a Ph.D. in pathology from the University of Oregon. He directs the program in toxicology at the University of Maryland and is a professor in the Department of Pathology at the University of Maryland Medical School. Prior to 1987, he held the positions of senior staff fellow and research biologist at the National Institute of Environmental Health Sciences in North Carolina. He chaired the NRC Committee on Measuring Lead in Critical Populations and was a member of the Toxicology, Women in Science and Engineering, and the Biological Markers of Urinary Toxicology committees. He chaired the Maryland Governor's Council on Toxic Substances. He was Fulbright Fellow to the Karolinska Institute in 1994 and is the winner of the 1998 Society of Toxicology Colgate-Palmolive Visiting Professorship to the University of Washington in *in vitro* toxicology. His research interests include the ultrastructural/biochemical characterization of mechanisms of cell injury from exposure to trace metals in mammals and marine organisms in relation to intracellular binding of both toxic and essential metals; biochemical mechanisms of metal-induced alterations of cellular heme metabolism; and molecular mechanisms of metal-induced alterations in gene expression.

HERSCHEL E. GRIFFIN received a B.A. from Stanford University and an M.D. from the University of California Medical School, with an internship in surgery at the University of California Hospital, residence in surgery at San Francisco Hospital, and subsequent graduate work in preventive medicine and public policy. He is professor emeritus of epidemiology at the Graduate School of Public Health, San Diego State University. Prior to his university appointment, Dr. Griffin was a practicing surgeon with the U.S. Army and was chief of the Communicable Disease Branch and the Preventive Medicine Division of the Office of the Surgeon General, among other posts. He also had a private medical practice. He has served on numerous health advisory boards and is presently on the California Potable Reuse Committee, established to advise the State of California. He has served on several NRC committees and is a past member of the NRC's Board on Toxicology and Environmental Health Hazards.

CHARLES N. HAAS earned a B.S. in biology and an M.S. in environmental engineering from the Illinois Institute of Technology and a Ph.D. in environmental engineering from the University of Illinois. He is the Betz Chair Professor of Environmental Engineering at Drexel University. He was formerly a professor and acting chair in the Department of Environmental Engineering at the Illinois Institute of Technology. His areas of research involve microbial and chemical risk assessment, hazardous waste processing, industrial wastewater treatment, waste recovery, and modeling wastewater disinfection and chemical fate and transport. He has chaired a number of professional conferences and workshops, has served as a member of several advisory panels to the Environmental Protection Agency, and is currently on an advisory committee to the Philadelphia Department of Health.

CHRISTINE L. MOE earned a B.A. in biology from Swarthmore College and M.S. and Ph.D. degrees in environmental sciences and engineering from the University of North Carolina. She is an assistant professor in the Department of Epidemiology, University of North Carolina, and an adjunct assistant professor in the Division of Environmental and Occupational Health at the Emory University School of Public Health. Prior to that, she was a fellow in the Division of Viral and Rickettsial Diseases at the Centers for Disease Control and Prevention in Atlanta. Her areas of research involve environmental transmission of infectious agents, especially waterborne disease, and the diagnosis and epidemiology of enteric viral infections.

JOAN B. ROSE earned an M.S. from the University of Wyoming and a Ph.D. in microbiology from the University of Arizona. She is currently a professor in the Marine Science Department at the University of South Florida. Prior to holding that position, she was a professor in the Environmental and Occupational Health Department at the University of South Florida. Dr. Rose is a fellow of the American Academy of Microbiology and past president of the Florida Environment Health Association. Her research has focused on methods for detection of bacteria, viruses, and parasites in wastewater, drinking water, and the environment, as well as on risk assessment and water treatment for removal of pathogens. She served on the NRC's Committee on Wastewater Management for Coastal Urban Areas.

R. RHODES TRUSSELL earned a B.S. in civil engineering and M.S. and Ph.D. degrees in sanitary engineering from the University of California, Berkeley. He is the lead drinking water technologist for Montgomery Watson, Inc. He has been involved with the evaluation of water reuse alternatives in a number of settings. He was involved in designing one of California's first major water reclamation facilities (Water Factory 21) and currently directs the San Diego Indirect Potable Reuse Project. Dr. Trussell serves on the Environmental Protection Agency Science Advisory Board's Committee on Drinking Water. He has served on a number of NRC committees and is a member of the National Academy of Engineering.

STAFF

JACQUELINE A. MACDONALD is associate director of the National Research Council's Water Science and Technology Board. She directed the studies that led to the reports *Innovations in Ground Water and Soil Cleanup; Alternatives for Ground Water Cleanup; In Situ Bioremediation: When Does It Work?; Safe Water From Every Tap: Improving Water Service to Small Communities;* and *Freshwater Ecosystems: Revitalizing Educational Programs in Limnology.* She received the 1996 National Research Council Award for Distinguished Service. Ms. MacDonald earned an M.S. degree in environmental science in civil engineering from the University of Illinois, where she received a university graduate fellowship and Avery Brundage scholarship, and a B.S. degree magna cum laude in mathematics from Bryn Mawr College.

GARY KRAUSS directed this study as staff officer at the National Research Council's Water Science and Technology Board until July 1997. He received his B.S. in zoology from Drew University and M.S. in ecology from the Pennsylvania State University. He directed the studies that led to the reports *Mexico City's Water Supply; The Use of Treated Municipal*

Wastewater Effluents and Sludge in the Production of Crops for Human Consumption; and *Water and Sanitation Services for Megacities in the Developing World.*

ELLEN A. DE GUZMAN is a senior project assistant at the National Research Council's Water Science and Technology Board. She received a B.A. from the University of the Philippines and is currently taking classes in economics at the University of Maryland University College.

EDITOR

DAVID A. DOBBS is a freelance writer and editor specializing in environmental, building, health, and science issues. His book *The Northern Forest*, coauthored with Richard Ober, won the Vermont Book Publishers Association Book of the Year award in 1995. Dobbs has edited texts on sports physiology, construction, sailing, horticulture, and natural resources issues; he writes frequently for a variety of publications including *Popular Science, Forest Notes, Boston Globe, Vermont, Vermont Life, Parenting,* and *Eating Well.* Originally from Texas, he received his B.A. degree from Oberlin College and now lives in Montpelier, Vermont.

Index

A

Abortions, spontaneous, 193
Activated carbon. *See* Granular activated
    carbon (GAC)
Activated sludge process, 21, 25, 91-92,
    104
Acute gastrointestinal illness (AGI), reports
    of, 77, 83
Adenovirus, 80-81, 85
Adsorption/desorption processes, 61-63
Advanced wastewater treatment (AWT)
    plants, 25, 30, 47-50, 53, 68, 241. *See*
    also Wastewater; Water treatment;
    Water treatment facilities
    emergency response plans at, 52
Advanced-oxidation processes. *See*
    Oxidation processes
Aerated lagoons, 91, 97
Aeromonas, 87-88
Agencies. *See* Regulatory issues; Water
    agencies
Agricultural chemicals, 4-5
AIDS patients, chronic diarrhea in, 87
Air stripping processes, 49, 236
Algae control, 25
Alkylphenol ethoxylate, 180
Alkylphenoxy ethoxycarboxylates
    (APDCs), 64

Alkylphenoxypolyethoxylates (AP*n*EOs),
    64-67
Alkylphenylpolyethoxycarboxylates
    (AP*n*ECs), 66-67
Alternative disposal routes. *See* Water
    disposal
Ambient water
    microbial contaminants in, 92-93
    mixing reclaimed water with, 2
Amebic dysentery, 79
American Public Health Association, 229
American Water Works Association, 209
Ames test, 138, 165, 169-172
Analytical tools, developing new, 61-63
Anthropogenic contaminants, 4, 17-19, 227
Aquatic humus, 4
Aquifers, injecting wastewater into, 2, 4, 22,
    24
Arizona. *See also* Salt River Valley
    Department of Environmental Quality
        (ADEQ), 41-42
    Department of Water Resources
        (ADWR), 41-42
    water reuse regulations in, 31, 41-42, 96-
        97, 120
Artificial recharge. *See* Ground water
Assessment. *See* Health and safety testing;
    Risk assessment
Astrovirus, 80-81, 84-85

253